PARTNERSHIPS FOR GLOBAL DEVELOPMENT

THE CLEARING HORIZON

DECEMBER 1992

A Report of the

CARNEGIE COMMISSION
ON SCIENCE, TECHNOLOGY, AND GOVERNMENT

® UNION GCIU LABEL 196-C

ISBN 1-881054-04-7

Printed in the United States of America

CONTENTS

FOREWORD

The familiar division of our planet into Western and Eastern Blocs and First, Second, and Third Worlds has vanished suddenly. But as the curtains that divided us and distorted our social, economic, and military relations have been pulled back, the scope of human needs has become much more visible. This report, prepared by the Task Force on Development Organizations, presents the compelling logic for renewed global partnerships in development.

The report emphasizes that science and technology are among the most powerful tools for development, especially allied with democratic values, market-friendly economics, and human rights. The Task Force argues that the United States, the world leader in science and technology, should rededicate itself to global development.

Such rededication involves not only a logic for action, but also an agenda and means. This report lays out the principles and criteria for the agenda and provides examples. It then identifies barriers that need to be removed and explains how organization and decision making should be changed to make genuine partnerships for development even more effective.

We are grateful to former President Carter for his leadership of this effort, visionary yet pragmatic. And we are deeply impressed that the members of the Task Force, representing extraordinarily diverse views and experiences, could find so much common ground. The members of the Task Force are:

President Jimmy Carter (Chair)
Rodney W. Nichols (Vice Chair)
Anne L. Armstrong*
Harvey Brooks
John R. Evans
Robert W. Kates
John P. Lewis
Lydia P. Makhubu
M. Peter McPherson
Rutherford M. Poats
Francisco R. Sagasti

George P. Shultz (senior advisor)

The report offers a remarkable basis for the government and people of the United States to move ahead, in the words of the report, "to make the world economy work for everyone and to help provide for those for whom the economy currently does not."

William T. Golden, Co-Chair
Joshua Lederberg, Co-Chair

* Participated as a member of the Task Force from 1990 through July 1992 and endorses the general conclusions and recommendations.

PREFACE

The Carnegie Commission on Science, Technology, and Government established the Task Force on Development Organizations to examine the changing circumstances facing cooperation for international development over the next decades and to assess the policy and organizational implications of these changes for the United States. The Task Force was chartered to address three questions: *Why* should there be cooperation for development? *What* should it consist of? *How* should it be organized? The report sets forth the argument for action and offers specific recommendations for the substance of programs and for ways of carrying them out.

The Commission was aware in establishing the Task Force that there are many reports on cooperation for development, and partly for this reason an extensive bibliography is included. There have been many calls to action. And there have been many pleas for programs to address particular facets of development, including agriculture, population, manufacturing industries, environment, health, basic human needs, gender issues, and children.

The Commission established the Task Force in the belief that a balanced report on the issues would be valuable, including *criteria* for the selection of programs and the often-neglected dimensions of organization and decision making, which are the Commission's special concern in all its work. In fact, the Commission felt that it could hardly consider its work complete unless it took account of the portions of humanity who are most in need of the benefits of science and technology.

The Commission is carrying out several activities aimed at strengthening the institutions and decision-making processes by which the use of science and technology is connected to world affairs. These activities are overseen by an International Steering Group chaired by Rodney W. Nichols. The Commission has issued a report on *Science and Technology in U.S. International Affairs* (January 1992), which examines how the U.S. Government, particularly the State Department, can be better prepared to mesh science and diplomacy. It has also issued *International Environmental Research and Assessment* (July 1992), which seeks to renew a positive, long-range vision of the international institutional science and technology infrastructure as it relates to environment and development. In addition, it has sponsored a consultant report by Alexander Keynan on "The United States as a Partner in International Scientific and Technological Cooperation: Some Perspectives from Across the Atlantic" (June 1991).

The Task Force on Development Organizations was created in early 1990, following a preparatory workshop sponsored jointly by the Commission and the Council on Foreign Relations and held at the Carter Center in Atlanta in October of 1989. Rodney Nichols, Vice Chair of the Task Force, and Jesse Ausubel, the Commission's Director of Studies, organized the workshop and guided the research and analyses that underlie the report. The Task Force met four times and also convened a session with the leaders of major private voluntary organizations involved in development. Members of the Task Force benefited from numerous other consultations and extensive correspondence with knowledgeable individuals, including leaders of industry and of governmental and intergovernmental foreign assistance programs, workers in the field, and scholars in development. Appendix A provides a list of those with whom we have worked.

Susan Raymond was the principal consultant to the study, and she and Charles Weiss prepared a series of particularly valuable background papers, which are available from the Commission and are listed in the Bibliography. William Foege, Executive Director of the Carter Center, and Victor Rabinowitch and Walter Rosenblith, members of the Commission's International Steering Group, participated extensively in the deliberations of the Task Force. Kenneth Keller, John Temple Swing, and Peter Tarnoff worked with the Task Force on behalf of the Council on Foreign Relations, and John Blackton provided liaison with the U.S. Agency for International Development.

Maryann Roper, assistant to President Carter, provided professional and administrative liaison with the Carter Center. James Brasher, Michael Heisler, and Nancy Konigsmark helped greatly at the Carter Center. David Kirsch, Margret Holland, Doris Manville, and Georganne Brown made many practical contributions to the success of the project, and the Commission's executive director, David Z. Robinson, offered valuable suggestions and consistent encouragement throughout the effort. The report was edited by Jeannette Lindsay Aspden.

The report is endorsed by the Task Force and was approved by the Commission at its June 1992 meeting.

INTRODUCTION

It is a privilege to introduce this rare and special report. It addresses a badly neglected and critically important topic. If these problems cannot be solved in the next few decades, there will be immense worldwide dangers.

When I set out to establish the Carnegie Commission on Science, Technology, and Government, with the superb cooperation of the Carnegie Corporation board, one of our deepest aspirations was to make this enterprise a truly international one, taking account of the extraordinary circumstances of world transformation. At that time, we were just beginning to envision the end of the Cold War and a vast array of new opportunities that might well emerge.

A vital turning point in the development of the Carnegie Commission on Science, Technology, and Government occurred when I was able to enlist the distinguished leadership of Joshua Lederberg and William Golden as co-chairs, and then to enlist the active participation of President Jimmy Carter in all the activities of the Commission. In light of President Carter's profound commitment to developing countries, including the uses of science and technology for development, the commitment to human rights and the building of democratic institutions, it was our privilege to ask him to chair the Task Force on Development Organizations. His agreement to do so, and his vigorous intellectual leadership of the enterprise, assured its success. We were doubly blessed by being able to link President Carter with an established leader of the scientific community who has had extensive experience in international cooperative efforts, Rodney Nichols. Their combined leadership has stimulated extraordinary cooperation from a diverse, broadly informed task force that is deeply international, interdisciplinary, and rich in relevant experience. Their approach has been open-minded, inquiring, and innovative. The result is a major statement—in my view, one of the most significant ever made on this subject, especially in view of its timing, context, and leadership. Its key points surely deserve emphasis at the outset.

The Cold War is over. With its grotesque distortions removed, new opportunities emerge throughout the world. So it is a time for basic reassessment, similar to the period right after World War II. For a variety of reasons, many nations in the Southern Hemisphere have been late in getting access to the remarkable opportunities now available for economic and social development. They are seeking ways to modernize in keeping with their own cultural traditions and distinctive settings. How can they adapt useful tools for their own development?

- The immense power of science and technology can be brought to bear on development throughout the world.
- The best context for bringing this about is the creation of democratic societies that strongly protect human rights.
- Market-oriented economies are vital in the modernization process.
- The application of these modernizing trends must be carried out in the framework of humane, compassionate values, and a concern for social justice.
- Such efforts will be pursued most effectively with a strong cooperative outlook and a mutual aid ethic as well as institutions to match—that is, cooperative global development.
- The United States can make a major contribution in fulfilling the promise of this approach, working in concert with many others—not dominating but stimulating and cooperating in international efforts.
- All this can best be achieved with a strong commitment to pluralism both within nations and among them. Pluralistic activity in public, private, and independent sectors can lead to vitality and creativity, with protections provided by the balance of such pluralism.

■ Altogether, the essential ingredients for development center around knowledge, skill, and freedom. Knowledge is mainly generated by research and development; skills are mainly generated by education and training; freedom is mainly generated by democratic institutions.

■ The orientation of cooperative international development is deliberately inclusive. The international community can find ways to work cooperatively for the benefit of most, if not all, nations.

■ The United States needs to create a national roundtable for international development; it also needs new legislation to establish much stronger aid institutions, policies, and practices—using many sectors to do so.

■ There is a great opportunity to strengthen multilateral organizations so that they can be more effective in using science and technology for development.

■ Both bilateral and multilateral activities must now facilitate productive civilian economies and civil societies; they must diminish the overwhelming commitment to the military sector that characterized the years of the Cold War.

Such a sketch cannot do justice to an intellectually rich and morally uplifting report. The main themes of this report are likely to reverberate throughout the world for decades to come. If they are taken seriously by leaders and incorporated into the work of relevant institutions, the world will become a much better place than it is now.

David A. Hamburg
President
Carnegie Corporation
of New York

I

THE CASE AND THE RECOMMENDATIONS IN BRIEF

TIME FOR RENEWAL IN A WORLD OF CHANGE

It is long past time to renew the content and form of the relations between the United States and the diverse countries loosely called "the developing world." Many of these countries, late to develop modern economies, are home to hundreds of millions of people still painfully burdened with illness and poverty. Many others have made great progress over the past generation. Moreover, there is a conjuncture today of welcome geopolitical change with a worldwide move toward market economies.

Although the situation is fluid, and hazards and reversals are all too obvious, these changes offer enormous opportunities. The world could move from merely preserving an armed truce, with bitter ideological tension, toward achieving peace, democracy, sustainable economic growth, and improvement in the quality of life. Yet in the United States existing laws and apparatus for "assistance"—or better, for cooperation for development—

are outdated. Even the will to support programs abroad may sometimes seem lacking. So the United States must define a new strategy, with firmer criteria, to govern the choice of its programs and its investments for international development. In renewing both the national commitment and the governmental organizations to pursue that strategy, the theme should be "Partnerships for Global Development."

Some say that the failures of the United States to ensure equality of opportunity and a decent standard of living for everyone calls into question America's right to prescribe societal changes elsewhere. And some argue that scarce resources should be applied to resolution of domestic problems, rather than devoted to "foreign aid"—charity, after all, begins at home. Yet improving economic (and, indeed, political) conditions throughout the world is not only correct, but necessary—it is ethical self-interest grounded in the principles of political and economic liberty endorsed by the United Nations and by free peoples everywhere.

What is advocated here is not outmoded "foreign aid" but modern partnerships for global development. As peace and prosperity spread throughout the parts of the world now crippled by unrest and poverty, economic opportunities for America will increase, and the reduction in international tension will mean that every nation can turn to its own pressing domestic needs. The United States has a chance not only to do good but to do well, to foster independence everywhere and to reestablish leadership for durable interdependence.

Still, any report on the controversial subject of "foreign assistance" must confront the three classical questions: Why?—is the rationale sound? What?—do the programs make the best sense? How?—is the implementation effective? The report's findings and conclusions answer these questions. This summary gives the case in brief, along with recommendations for action.

THE CASE

In the National Interest

The United States has compelling interests—a mix of humanitarian, economic, and security reasons—to promote cooperation for development. American goals in health, environment, jobs, exports, and conflict resolution are all interdependent with the actions of others around the world. Moreover, as the world's most powerful nation—with a tradition of generosity as well as leadership in science and technology—the United States brings unique assets to a partnership for development.

A basic principle of American cooperation will remain: to foster everywhere the balanced development of the private, public, and independent sectors. The resulting pluralism nurtures diversity and encourages constructive competition that will test and improve even the best ideas. Cooperation of this kind can be achieved only by broader and better-balanced participation of the different sectors in the United States and in all donor countries. The overall U.S. aim in successful partnerships with every country is the establishment of sound public administration, a culture of lively enterprises, a healthy not-for-profit independent sector, and a shared commitment to political freedom, social opportunity, and unfettered worldwide trade.

Science and Technology: A Key to the Future

For the 1990s on into the 21st century, science and technology will continue to be a linchpin in the efforts to achieve most of the world's social and economic goals. They undergird the research that creates needed knowledge. They help build the education and training systems that advance skills. And they thrive with the freedoms of inquiry, communication, and association that ensure, and are ensured by, democracy and liberty.

Program Criteria and Illustrations

What programs shall the partnerships select? Along with intrinsic merit, there are four criteria for the selection, design, and conduct of programs in any country. One criterion is the policy environment, especially economic trends. A second is the prospect for ecological and social sustainability. A third criterion is the potential to build human and institutional capability to solve future problems. Fourth, partnerships need sturdy lines of communication to promote the social understanding that enables the establishment of mutual objectives and shared responsibilities.

As these criteria are applied, initiatives of immediate importance to the alleviation of desperate human suffering must be pursued, employing what we already know. At the same time, a longer-term outlook must challenge the vast potential of science and technology to discover better means for accelerating social advance by applying new ideas.

Many goals demand urgent application of the potential of science and technology: halving world hunger, reducing the incidence and the toll of tuberculosis, protecting and restoring the earth's forests where they are at risk, building the capacity for economic policymaking in the nations of the so-called Third World, and relating the U.S. educational enterprise to the needs and aims of development. Programs are needed in all these areas, and more; and programs already established — some many years ago — must be updated to make best use of science and technology for development.

U.S. Action

How shall the United States proceed? Renewed U.S. cooperation in global development will require a significant strengthening of national and governmental capacity and willingness to work with the full spectrum of developing countries, from the poorest to the newly industrialized. Moreover, the United States must enhance its efforts to help solve problems that cut across national borders, notably in health and the environment.

Programs must be driven by needs in the field. They must be freed from outdated objectives as well as from the obsolete political, economic, and geographic constraints that in the past determined eligibility for action and funding. Cooperative development must establish a more effective balance between growth and equity, management and participation, large-scale and small-scale endeavors, global campaigns and local needs, and the establishment of rules and norms and investment in bricks and mortar. Technological savvy—an awareness of what might work, and an analysis of why and how—will be essential for almost every program.

An imperative for implementing the next generation's partnerships in global development is that the United States must harness much more fully the power of its own pluralism. Government at the federal, state, and city level—along with the private for-profit sector—must reach out to the independent sector, including private voluntary organizations, universities, and foundations. All must improve their ability to work together across institutional lines, forming coalitions to press ahead on the actions needed internationally.

THE RECOMMENDATIONS

The recommendations touch upon every area of activity of the United States and illuminate the many new ways in which action must be taken in international partnerships.

National Action

■ **To foster creative cooperation among all U.S. institutions, a National Action Roundtable for International Development should be created, with balanced representation from the private, governmental, and independent sectors** (see pages 84–85). The purpose of the Action Roundtable would be to review the evidence on trends and then catalyze the formation of specific task forces to address urgent problems. Some task forces would focus on a particular nation or region, others on a technological opportunity, and still others on a longer-range process such as educational institution building—every group proceeding on a specific plan and timetable. Each action would be clearly in the international interest, and each would need to be justified in a convincing way to the American public.

Multilateral Action

■ **The United States should encourage and take a leading role in an analysis of multilateral organizations to identify opportunities to improve their performance, frequently by using science and technology more perceptively** (see pages 102–103). As critical as change in the national strategy is, a change in outlook on the world is also crucial. In short, multilateral action is often the best way to solve global problems.

■ **Greatly enhanced means must be devised for coordinating the ongoing efforts of the major donors** (see page 102). Such coordination would be aimed at achieving better results, given the changing circumstances in the field. Special attention should be given to the international capacity for studies and research on the most difficult and longest-range problems in science and on technology pertinent to development: new institutions may be needed. The increased emphasis on multilateral work and enhancing donor coordination will by no means eliminate the vital roles for bilateral programs.

White House and Congressional Action

The most important recommendations for the federal government are directed at the highest levels of the Executive and Legislative branches.

■ **The White House must take the lead** (see pages 85 and 91–92). Entrenched interests, institutional inertia, and organizational complexity—developed over more than forty years—require the President to articulate anew the principles and long-range priorities for cooperation with the entire range of developing countries. A bipartisan outlook will be essential. Presidential guidance should draw upon an intensive review by all relevant federal agencies of their current and desired activities with and in developing countries. To be completed during 1993, this complex review must be started now.

■ **Concurrent with new presidential leadership, the Congress should initiate broad consultations, studies, and hearings that will lead to major reform of "foreign assistance" legislation and oversight** (see pages 86–90). Given the public's skepticism about "foreign aid," and the many domestic urgencies, the political problems in Congress are exceedingly difficult. Yet in recent years, sweeping and constructive changes have been outlined by congressional and executive leaders of both parties. These new paths must be taken. At a minimum, the reforms include setting only a few broad goals, imposing much less detailed constraints on programs and funds, and relating global development strategy to foreign policy aims while keeping U.S. economic and social goals in sharp focus.

Independent Sector

■ **Leading organizations in the independent sector concerned with partnerships for development using science and technology should explore new mechanisms for regular exchange of information and extension of voluntary networks to address common concerns** (see pages 82–83). The mechanisms should be sharply problem-oriented so that participants see their shared mission and fulfill action-plans. Although universities, foundations, and many nonprofit centers have extraordinary competence, their effort has been fragmented, and it has not been shored up with long-range research.

Private Sector—Business, Labor, and Industry

■ **Major private-sector organizations should form study groups and action-oriented panels on the key issues in international development** (see pages 83–84). The point is to link high-level U.S. business executives for exchanges of ideas about economic policy, both domestic and international, concerned with long-range global development. U.S. private enterprise and labor must recognize and act to realize the benefits of trade with developing countries and the rewards of the accelerated global economic growth that will accompany cooperation for development. The proposed National Action Roundtable should facilitate communications with the independent sector and with government so that the private sector can become more broadly engaged.

Executive Agencies

■ **The means for interagency program development must be strengthened** (see page 92). Many federal departments and agencies with science and technology capabilities participate in foreign projects, but there is much too little coordination for development across agencies.

■ **To fulfill its mandate, the Agency for International Development (AID) must increase its access to American expertise in science and technology, enhance staff skills, decentralize authority, improve long-range planning, and match its organization to evolving international conditions** (see pages 92–97). AID is the U.S. Government organization with the most significant explicit financial and policy responsibility for "foreign assistance." Although presidential leadership and legislative reform will have to precede AID renewal, recent appraisals of AID have made abundantly clear what must be done—and the task, while difficult, is feasible.

RESOURCES

■ **The United States can afford to—and should—rededicate itself to a fair share of the effort on urgent development in Africa, Latin America, Asia, and the Middle East *and*, at the same time, reach out to the extraordinary opportunities in Eastern Europe and the former Soviet Union** (see pages 103–105). Even with constrained national budgets in the United States (and in other donor nations), surely there should be plans for a shift in "aid" budgets from military to development purposes. In parallel, the developing countries should shift their expenditures from military to civil accounts. Furthermore, these shifts of public resources must be integrated into the vastly larger context of the flows of private savings and investments throughout the world. Overall, new strategies must place public funds within a framework that enhances private incentives for economic growth.

CREATIVE COMMITMENT NOW

It is time to break away from the obsolete images of the world of the 1960s, the 1970s, and the 1980s. That world no longer exists. Now is a rare moment, a clearing horizon of historic opportunity, for all nations to promote peace, liberty, and global prosperity through partnerships. It is a unique time for creativity, comparable to the era immediately after World War II. Concepts, laws, and institutions must change. The stakes are high. So are the chances for success. It is time for the United States to use its human and financial resources to make the world economy work for everyone—and to help provide for those for whom the world economy currently does not. For many reasons—humanitarian, economic, and security—this is, indeed, profoundly in the national interest.

2

WHY: THE REASONS FOR COOPERATION FOR DEVELOPMENT

The central goal of development is the realization of the full potential of all individuals in their societies. Without compromising options for future generations, this realization of potential should enlarge the range of people's choices and make development more democratic and participatory. Choices should include access to income and employment, education, health, and a clean and safe environment.

This chapter reviews the roots of cooperation for development, explores dimensions of change as well as enduring continuities, and discusses the particular interests and assets of the United States in development, showing why now is the time for action.

ROOTS OF CURRENT COOPERATION

So much about today's world has changed so rapidly that it often seems that even the foundations of international relations have shifted. Much the

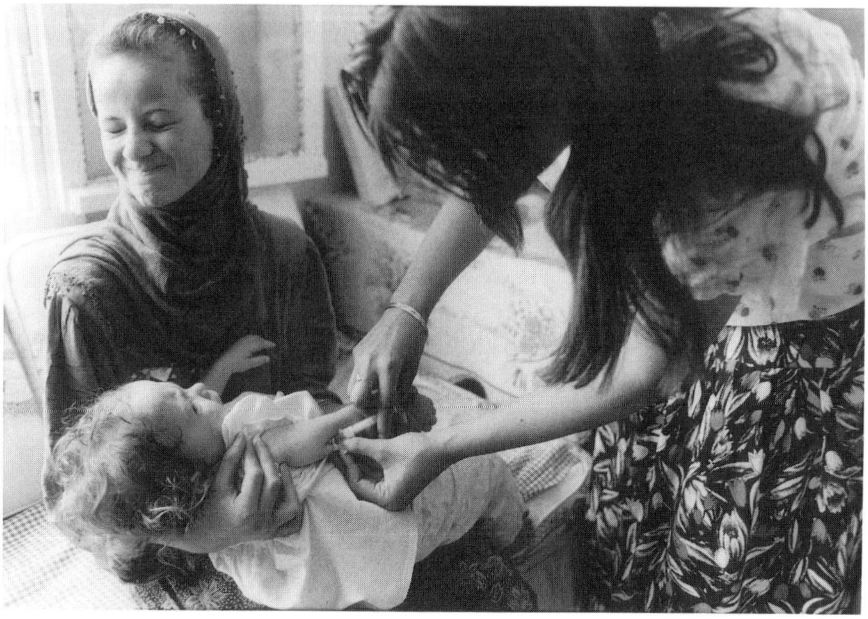

Immunization against diphtheria, whooping cough, and tetanus, Adana, Turkey.
(Photograph by Christopher Warren.)

same can be said about the relations between the United States and those countries late to develop modern economies and home to many hundreds of millions of people still painfully burdened with ill health and poverty. To understand where these relations now stand, and where they will and must go tomorrow, a brief historical sketch is instructive.

The efforts of the United States to provide assistance to other nations are rooted in economics and politics as well as in the humanitarian spirit that American culture manifests.

Economics

From the outset, a powerful motivation for foreign assistance has been economic. As can be seen in Box 1, even before the end of World War II, enlightened Allied statesmen recognized that their self-interest lay in rapid and widespread economic recovery. The appreciation that flourishing economies are mutually reinforcing led directly to the programs for rebuilding Europe and Japan. It generated the effort to stabilize currencies and to ease and expand trade through the creation of the International Monetary Fund and the General Agreement on Tariffs and Trade (GATT). It brought forth the World Bank to provide guarantees and investments for reconstruction and development. And the United States led these multilateral efforts while at the same time beginning new bilateral programs.

Geopolitics and the Birth of The Third World

Almost simultaneously, a second concern, geopolitics, also spurred foreign assistance. Expanding Soviet influence evoked the policy of "containment" and an East–West axis of political and military confrontation. The leading institutions of this confrontation were military organizations, especially the North Atlantic Treaty Organization (NATO) and the Warsaw Pact. They sought to protect territory and deter war through a balance of fear that would eventually stretch from Europe to Southeast Asia, Africa, and Latin America. Through these organizations a global drama was played out on a stark and forbidding stage furnished with rapidly expanding arsenals of nuclear weapons. Members of each alliance fostered the economic growth of their own side through the Organization for Economic Cooperation and Development (OECD) in the West and the Council for Mutual Economic Assistance (COMECON) in the East. Sometimes each alliance sought to hinder the economic progress of the other.

In this simmering Cold War, competition for allies was keen, especially within the "Third World," as it was first called by Alfred Sauvy in 1952.[1] As developing countries achieved independence, they were avidly courted by the opposing camps. "Development assistance" became the main currency in the struggle for loyalties along the emerging North–South axis of the international economy. It brought both badly needed finance and widely desired military might. Both East and West promised the poor nations of the South that theirs was the fast, reliable, and politically correct track to industrialization and wealth.

The Desire To Alleviate Suffering

A third stream of motivation flowed between the North and the South in this period. The interest in development in the North was nourished by a genuine desire to support new nations seeking to grow from colonialism to self-reliance. Recognition of these newly independent nations' aspirations for progress and betterment, as well as the deep poverty in which those aspirations were mired, elicited the humanitarianism that underlies public

Box 1. Chronology of Organization for Development[a]

1944	Bretton Woods Conference to plan for postwar Europe and to organize recovery assistance
1945	Creation of the International Bank for Reconstruction and Development (the World Bank) for long-term capital lending
1946	Creation of the International Monetary Fund to correct short-term financial imbalances
1947	General Agreement on Tariffs and Trade (GATT) developed as Executive Agreement within proposal for an International Trade Organization (ITO)
1947	U.S. bilateral European Recovery Program (Marshall Plan) authorized
1948	Creation of the Economic Cooperation Agency (ECA) to administer the Marshall Plan; ECA later renamed the Mutual Security Agency (MSA)
1949	Point Four Program announced to expand bilateral aid to developing countries
1950	Creation of the Technical Cooperation Administration (TCA) to implement the Point Four program
1953	Foreign Operations Administration (FOA) formed to combine MSA and TCA
1954	Mutual Security Act passed to recodify all foreign assistance
1954	PL 480 program of food assistance created
1958	Development Loan Fund formed for U.S. capital assistance
1960	Export–Import Bank established as independent federal agency to ease export financing of U.S. goods and services
1961	Peace Corps established as independent federal agency for unsalaried American volunteers to work in villages in developing countries
1961	Foreign Assistance Act separated military from economic assistance, unified all economic assistance under a new Agency for International Development (AID), and emphasized long-term planning and programming
1964	*Gardner Report* reviewed AID technical capabilities
1965	GATT Part IV on Trade and Development adopted
1969	*Hannah Report* reviewed AID from the perspective of the universities
1970	*Peterson Report* reviewed AID against increasing public criticism arising from the Vietnam War
1971	Overseas Private Investment Corporation separated from AID; provides insurance against political risks for U.S. private direct investments in developing countries

Box 1. *(continued)*

1973	"New Directions" amendments to 1961 Foreign Assistance Act targeted bilateral assistance on basic human needs in order to restore public support for aid
1973	Title XII added to the 1961 Foreign Assistance Act to strengthen U.S. university role in food and nutrition
1975	Funds for politically/militarily important countries separated from general development funds; creation of two "accounts" within AID, Economic Support Fund (ESF) and Development Assistance (DA)
1977	*Babb Report* reviewed AID and proposed field-focused organizational changes to implement the New Directions legislation
1978	Creation of the International Development Cooperation Agency to link all U.S. Government organizations engaged in development
1978	Proposal for creation of the Institute for Scientific and Technical Cooperation (ISTC) for research and technical assistance
1979	*Gordon Report* reviewed AID from within the bureaucracy as a reaction to implementation of the Babb Report; emphasized role of research
1980	Trade and Development Program established to increase U.S. exports to developing countries by financing project feasibility studies
1981	Formation in AID of S&T Bureau and Private Enterprise Bureau to act together in applying S&T to assistance programs
1983	*Carlucci Report* reviewed AID in context of declining support for foreign assistance yet rising threat from the Soviet Union; recommended integrating military and economic assistance
1983	National Endowment for Democracy created to encourage autonomous economic, political, social, and cultural institutions throughout the world as foundations of democracy and guarantors of individual rights and freedoms
1989	*Hamilton Report* (U.S. House of Representatives) recommending new foreign assistance legislation and an overhaul of the administrative organization
1989	*Woods Report* from AID reviewed the economic and social condition of the developing world and raised questions about the appropriate organizational response
1992	*Ferris Report*—the report of the President's Commission on the Management of AID Programs—offers Congress a plan for reforming AID

[a] The eight reports in italics are analyzed by Charles Weiss in "Lessons from Eight 'Reform Commissions' on the Organization of Science and Technology in U.S. Bilateral Assistance," a background paper prepared for the Task Force on Science, Technology, and Government.

support for development assistance. A desire to alleviate suffering has always been a foundation of assistance to developing countries and will remain so in the future.

INSTITUTIONAL CONSEQUENCES OF AMBIVALENCE

From this period, then, an ambivalence about cooperation for development grew within the public and its government. Political, economic, and military motivations were often pitted against humanitarian concerns. Even when not diametrically opposed, they remained, at best, uncomfortable companions.

The mix of objectives was reflected in the series of institutions and programs created by the United States Government to conduct cooperation for development (see Box 1). After World War II, several institutions for economic and technical cooperation and mutual security were created to administer what became known as the Marshall Plan. As emphasis shifted from Europe to the Third World, the mandate of "foreign assistance" was broadened through the "Point Four" program to the newly independent nations and other nonindustrialized countries. The Peace Corps, created in 1961, captured the interest of many people, who volunteered to work in these countries. In the same year, several diverse ongoing programs were pulled together by the Foreign Assistance Act to form the Agency for International Development (AID), the government's primary agency for foreign assistance. Much money continued to be spent on improvement of large physical infrastructures. In the early 1970s, a major revision of the content of development cooperation emphasized the "basic human needs" of the world's poorest. During the 1960s and 1970s, the Export–Import Bank and the Overseas Private Investment Corporation were created to encourage private trade and investment in developing countries.

Over the years, the U.S. Government's institutions for global development expanded and contracted with presidential and congressional commitment. Rhetoric and programs shifted in response to new development theories and approaches. Both substance and organization were studied and restudied as the roles and limits of bilateral, governmental action came to be better understood. Nonprofit voluntary organizations multiplied and came to shoulder a larger share of cooperation, especially with the poorest regions of the world. The multilateral World Bank expanded, and other multilateral institutions, both global and regional, were created and, indeed, now dispense more development resources than any bilateral agency. The U.S. share of net disbursements of loans and grants made on concessional financial terms declined. More countries, including China, joined the international system. Arms negotiations reduced the threat of world war.

Old Structures, New World

Despite the many fluctuations in style and slogans, the overall institutional and legal framework of the United States Government's cooperation for development has changed little since the early 1960s. But the world of the 1960s, the 1970s, and even the 1980s no longer exists.

DIMENSIONS OF CHANGE

As in any analysis of human events, separating the multiple and intertwined strands of change is difficult. For the sake of order, however, we address first those that are primarily political, then those relating to human development and economics, and finally those relating to science and technology and to environment and natural resources.

Politics

In a torrent of events, the ice of post–World War II politics has broken and been carried away. The nations of Central and Eastern Europe and the Baltic region are newly free. The Warsaw Pact, the antagonist that evoked the free world's political and military defense of Europe, has dissolved. The Soviet Union is gone, and the future of the commonwealth and member states that have succeeded it is unclear.

Amidst these changes, not all results have been beneficial. The rise of nationalism and ethnic rivalries bodes ill for a smooth transition from Communism to liberty, justice, and prosperity. Yet the world senses an unprecedented chance for a deeper international peace and a wider horizon for development.

Changes, however, emanate not simply from the demise of the Soviet Communist bloc. Simultaneous, related events have altered the political face of many parts of the world. Many changes have been fundamental and more rapid than could have been predicted even five years ago. Reforms in southern Africa are under way. Namibia is an independent nation, and

South Africa has begun to dismantle the prison of apartheid. Elsewhere in Africa democracy is advancing.

A remarkable number of countries throughout the world have conducted or scheduled free elections since 1989. In South and Central America and the Caribbean, most governments are now freely elected. Across Africa, free elections are taking place. In the newly democratic nations of Eastern and Central Europe and the former Soviet Union, freedom of political and economic choice is replacing tyranny. Even in the Middle East, decades of confrontation have given way, however tentatively, to new perspectives and a still fragile dialogue about the region's future.

The positive momentum of political changes will not eliminate surprises and setbacks. Nor will an improved international climate solve all national problems. Political, ethnic, and religious fragmentation and crisis continue to plague the Yugoslav region and could intensify in India. China's future remains uncertain, and its policies, cohesion, and economic well-being will seriously affect world events. Several scenarios in Southeast Asia are possible. The future of the planet is fluid, and, while the opportunities for positive change are many, the speed of events carries the risk that the expectations of newly free peoples will rise far too fast for governments and economies to satisfy them, setting the stage yet again for disillusionment and conflict.

Human Development and Economics

Under the turbulent surface of political change are incontrovertible facts of sustained material growth and improved welfare. The social, economic, and technological evolution of the past 40 years has been more consistent than the course of political evolution. It suggests a universal potential often forgotten amidst the dramas of revolutions, plagues, and floods that occupy the front pages each day.

Health, Education, and Well-Being

There have been real and significant improvements in many measures of health, education, and well-being around the world. Much of this improvement has been spurred by modern science and technology. In many cases and for many people, development has succeeded. For example, Third World infant mortality rates have declined rapidly (Figure 1). Progress that took England and France a century to achieve has been achieved in just decades in Latin America and Asia, and even in parts of Africa.[2]

Life expectancy is increasing (Figure 2). Between 1960 and 1988,

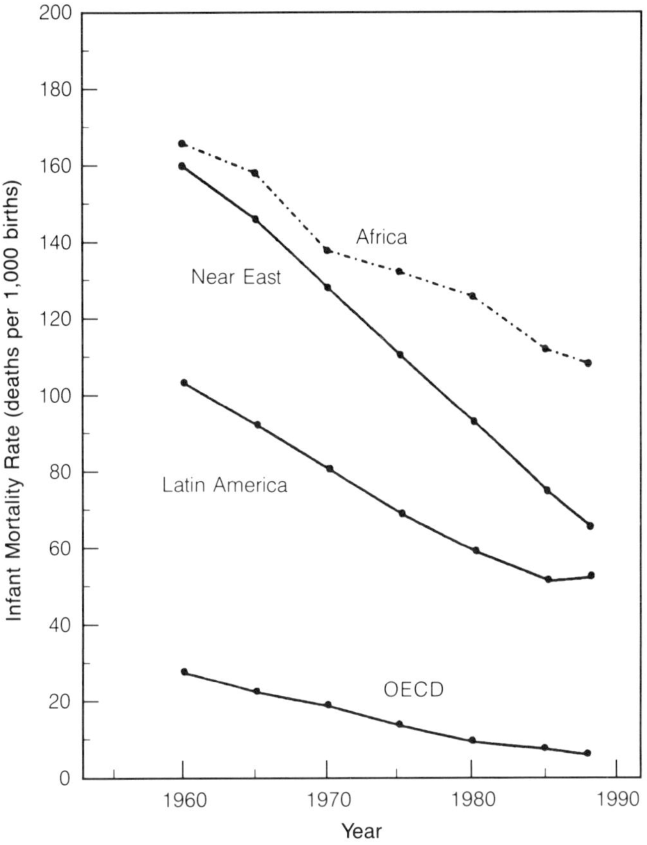

Figure 1. Changing infant mortality (deaths per 1,000 births) in three regions of developing countries and the OECD, 1960–1987. (Source of data: World Bank World Data Tables.)

life expectancies increased by 38 percent in the Near East/North Africa, 24 percent in sub-Saharan Africa, 23 percent in Asia, and 20 percent in Latin America.[3] Smallpox has been eradicated,[4] and rising rates of immunization are beginning to loosen the death grip of infectious diseases on many of the world's children.

Educational levels are also rising, and basic education is more widespread.[5] Even in low-income countries, the percentage of children enrolled in primary school has more than doubled (Figure 3).

There have also been improvements in nutrition. Only in sub-Saharan Africa does average per capita daily caloric intake generally fall short of the standard suggested by the United Nations.[6] Indeed, with the application of scientific advances and the Green Revolution, several former recipients

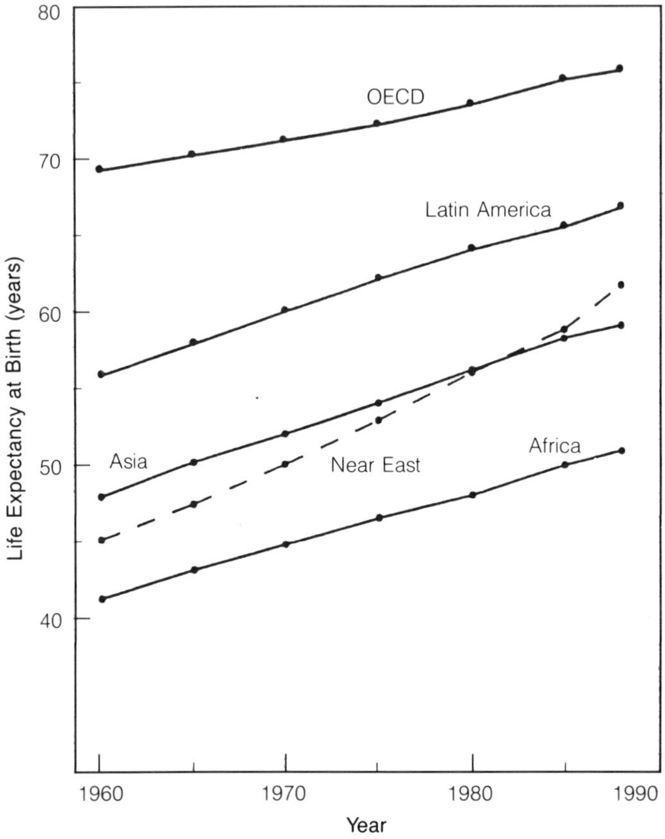

Figure 2. Changing life expectancy at birth in five world regions, 1960–1989. (Source of data: World Bank World Data Tables.)

of food aid now export food. Agricultural development has powered economic success in most countries. A billion people still go hungry, but this tragedy could be greatly lessened; reducing the number of hungry people in the world by half within the next decade is possible (see Box 3, page 58).[7] The most critical area of need remains sub-Saharan Africa, where agricultural self-sufficiency continues to be elusive and where averages mask the enormous grief in pockets of famine and malnutrition.

Economic Progress

Assessed in terms of economic output and income, the well-being of countries has also improved. On average, real per capita income in the devel-

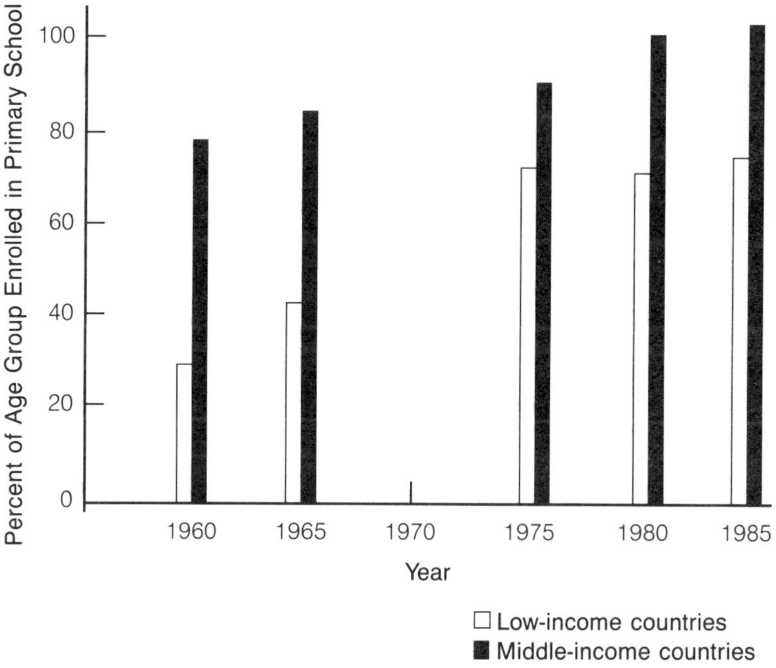

Figure 3. Percentage of school-age children enrolled in school in low- and middle-income developing countries. Total may exceed 100% because of older students enrolling in lower grade levels. (Source of data: World Bank, *World Development Report*, 1979–1991.)

oping world has grown over 2 percent per year since 1950.[8] In nearly all countries, the past four decades have witnessed economic progress,[9] although its benefits have often been uneven within societies. In some countries, progress has been remarkable. In East Asia, real annual per capita income growth of the "Four Tigers" (South Korea, Taiwan, Hong Kong, and Singapore) in the past decade has averaged 4.8 percent.[10] Improved well-being in the region is not limited to these economic engines. In the same period, Indonesia and Thailand were not far behind, with annual real growth rates approaching 3.5 percent[11] a rate that doubles income in 20 years.

However, even when economic growth is achieved with relative equity, recent history teaches again the bitter lesson that political and social stability are not guaranteed. Two of the historically better performers on the dimension of income equality, Yugoslavia and Sri Lanka, have become tragically entangled in destructive webs of ethnic strife.

Growth has been accompanied by and spurred by fundamental changes in economic structure. Developing countries are no longer simply sites where natural resources are mined. As agriculture becomes more productive and efficient, manufacturing and service sectors are growing and now provide more than 80 percent of the gross domestic product (GDP) of the South.[12] Among low-income countries, agriculture today accounts for only a third of gross domestic product, while in middle-income countries the share is down to 12 percent.[13] Over half of developing country exports are manufactured goods, up from 26 percent in 1965.[14] This increasingly diverse economic base means that more countries have greater prospects for stable, balanced growth. To capitalize on these opportunities in the future, much more widespread application of science and technology will be needed in the manufacturing and service sectors as well as in the creation of an educated, skilled workforce.

National economies now interact in a different world. The global economic anomaly created by World War II has passed. For two decades after the war, the world's economy was dominated by the United States. Almost 40 percent of gross world product originated in the United States, and its share of world exports soared.[15] The opportunity for leadership was often used well by the United States, as evidenced by its role in advancing trading institutions to create more open opportunities for commerce. This imbalance, however, also meant that U.S. institutions, both public and private, were unaccustomed to partnership.

By the mid-1960s, the U.S. share of global economic activity was returning to prewar levels as Europe and Japan rebuilt. The economic dominance of the United States receded, but U.S. agencies and businesses were often slow to recognize the rebirth of competition and to accept partnership roles. The development institutions of the United States and of the North in general retained their tendency to dictate development objectives and agendas.

Today, the balancing of the international economy has gone further. More nations participate, and participate more actively, in the international marketplace, and the weights of most of the players are more evenly matched. A greater share of all economic transactions is transnational. The globalization of finance has promoted rapid capital flows and wider access to the world's savings pool. Manufacturing crosses and recrosses national borders until the origin of a product is obscure; the making of parts and their assembly in stages take place in a variety of countries. Nations are increasingly interdependent, and no nation alone is in full control of its economic destiny.

Trade has expanded rapidly, entwining nations more closely. Since 1950, the value of trade has risen dramatically (see Figure 4).[16] The export of manufactured goods from developing countries more than tripled in value between 1970 and 1980, and has more than doubled again in the past decade.[17]

This growth in trade has created a world that is more economically

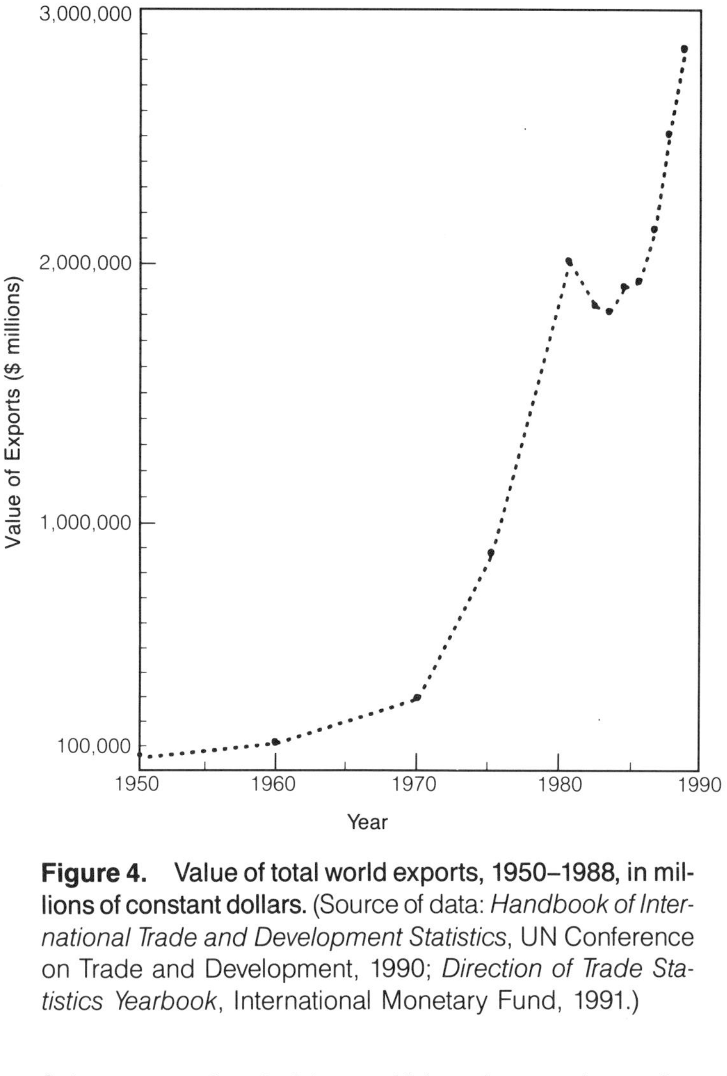

Figure 4. Value of total world exports, 1950–1988, in millions of constant dollars. (Source of data: *Handbook of International Trade and Development Statistics*, UN Conference on Trade and Development, 1990; *Direction of Trade Statistics Yearbook*, International Monetary Fund, 1991.)

integrated than at any time in history. Although exceptions exist and pressures for protectionism remain, another effect of this expansion has been a gradually broadening acceptance of the global benefits of free trade and the goal of further liberalization of trade regimes. The fate of the economy of the former Soviet Union is a chastening reminder of the withering long-run effects of isolation from the global market in information, goods, and services.

Science and Technology

Underlying the advances in well-being are progress in science and technology and the wide diffusion of innovations. Major improvements in plant breeding and irrigation during the last 40 years have enhanced the ability of many developing nations to feed themselves. Satellites used for weather prediction have helped mitigate the effects of natural disasters. Both low- and high-technology contraceptives have helped ensure that children that are born are wanted, and techniques such as oral rehydration have increased child survival. Information technologies have enabled new techniques for management and rapid data transfer in manufacturing and among financial institutions, shaping productive investments around the world. Electrification has brought light to villages, towns, and remote farmsteads alike. Modern communications and transportation link much of the developing world to the global economic and information network. Compared to 30 years ago, there are few national "islands" of isolation remaining among developing countries.

The adaptation of cultures to this ensemble of changes and the tides of internationalism, however, is fraught with uncertainty. In many places, longstanding institutions struggle to retain their traditional social roles. When their functions, such as care of the young and old, are not fulfilled, the human price is painfully obvious in the streets of any city in the developing world, and in the industrialized world as well. In this context, certain movements have emerged to confront and confound the pace and direction of economic, political, and social change, and the progress of science itself.

Environment and Natural Resources

Just as the political and social landscape has been altered, so has the physical landscape. In no other respect has the effect of enormous economic and population growth been so obvious or so serious. Tropical forests dwindle. Deserts encroach. Wetlands diminish. Species disappear. Lakes die. The ozone layer thins, and concentrations of greenhouse gases rise. Like modern-day Flying Dutchmen, vessels with hazardous wastes wander the seas in search of a port.

The state of the world's environment is being comprehensively documented, and concern for the future is widespread. One hundred and twenty heads of state gathered at the Earth Summit in June 1992 in Rio de Janeiro to promote improvement in environmental protection and conservation.[18] The evidence presses home to the public and governments alike that much more attention must be given to promoting and implementing policies and programs that support economic and social progress, while imposing a much lower price on the environment. The contributions of science and technology to reducing environmental threats today—and to developing less polluting technologies for tomorrow—have been and will be among the most significant factors in the Earth's environmental future.

The poor suffer most from a degraded environment. Research on environment and health risks shows that richer is safer and poverty is the worst toxin. The challenge thus remains to accelerate development, but in a way that damages less land, consumes less energy and materials, and delivers services more efficiently. Technologies are within reach that can make a large difference. But it is not yet certain whether they can be deployed fast enough to offset the environmental impacts of the economic growth needed to improve the lot of those who are poor now.

The imperative to ensure that economic and social developments are ecologically sustainable is newly appreciated in both industrialized and developing countries. In what might be called a "pollution equation," the high consumption of the North and the population growth of the South are weighted about equally. "Sustainability" is sure to prove to be one of the guiding principles of development and may provide the new glue for deeper, broader, and more productive global partnerships.

ENDURING CONTINUITIES

Even in the face of a powerful economic evolution and sudden political revolution, many development themes continue. Some of these continuities provide the foundation for future cooperation. Others are stubborn challenges.

NEED

The first continuity is suffering, human need, and untapped human potential. Although there have been many achievements, the vastness of what remains to be done is apparent in every region of the world, including the cities of the North. The hungry, uneducated, ill-clothed, and poorly housed outnumber the affluent in far too many places. A billion people throughout the world remain impoverished, fishing out a bare existence at the margins of the vast global resource flows. Aggregate figures hide large regional populations with persistently short life expectancies and high mortality rates. For example, although India's overall infant mortality rates have improved considerably over the last several decades, infant mortality levels among landless families in rural Punjab are 36 percent higher than those for landowning families.[19] And in Bombay, crude death rates in slum areas are as much as three times higher than in the city's suburbs.[20] Generally, there is evidence that the century's progress began to flatten out in recent years, especially in Africa and Latin America.[21] The 1980s can be called the "Lost Decade" of development in large areas.

Commitment and Resources

Continuing, but perhaps weakening, is the commitment to humanitarian programs that respond to the basic needs of the world's most disadvantaged. The strength of the national and international conscience undergirds cooperation for development. The total of private, voluntary contributions to overseas development by Americans—about $6 billion—is comparable to the U.S. federal budget for such assistance (and larger if expenditures by religious groups are included).[22] The $6 billion is an earnest expression of individual, human generosity and commitment that is frequently forgotten.

Governments also maintain a financial commitment to cooperation for development in both bilateral and multilateral forms. Measured in absolute terms, amounts are significant. Worldwide Official Development Assistance (ODA) totaled nearly $50 billion in 1990.[23] This is equivalent to the annual profits of the 50 largest industrial corporations in America.[24] In relative terms the amount is less impressive. It equals about $12 per person for a year for the population of all recipient countries. Since 1980 about $3.50 out of each $1,000 of GNP in the OECD countries has gone for official development assistance, half the target of $7 endorsed by the United Nations and supported in principle by most donor countries. Nevertheless, the traditional donor nations remain committed to cooperation. Furthermore, the nations whose incomes have significantly increased in recent years, whether by virtue of natural resource endowments (e.g., Saudi Arabia) or industrialization (e.g., Finland), have joined the ranks of development donors.

The public ethic that lies at the root of cooperation appears to have withstood the test of time. It was reaffirmed and extended at the 1992 Earth Summit. Whether out of self-interest or moral concern, the nations that succeed reach out to those still struggling on the path to success. Not only is there a larger number of potential partners in the North, but their economies have become more open. In fact, there is continuing growth of key groups of private institutions as partners within both developed and developing countries.

Many nongovernmental organizations and networks have expanded their global efforts in fields such as environment, health, and human rights. The number of U.S. private voluntary organizations with programs in the developing world tripled between 1973 and 1989, and U.S. Government funds allocated for their support increased seventeenfold.[25] Such institutions and their counterparts in the developing world will be increasingly able to design and conduct programs for economic progress and social welfare. The rise in this kind of public–private entrepreneurship offers an enormous resource now and an even greater potential for the future.

The North also has new and potentially more accessible technical capabilities. Universities and other research institutions have grown in size and number and have become more diverse. In the United States alone, well over a hundred universities now embrace substantial research programs, and university research expenditures increased by 75 percent between 1980 and 1990.[26] Universities have also become more international in character. Over half a million students from developing countries are enrolled in higher education in the United States, France, Germany, the United Kingdom, and other industrialized nations.[27]

A growing resource is the leadership within the developing world. In both the government and private sectors, a new generation of leaders is beginning to steer tomorrow's course. Many have themselves benefited from the advances of the last 30 years, and have the education and experience that encourages an openness to change.

Barriers

Because of, or perhaps in spite of, the speed and scope of change, other fundamental continuities exist. These are the barriers to world progress. Over the next decade, such barriers as poverty cycles, debt burdens, human rights abuses, military spending, unnecessary economic slumps, and underdeveloped science must all be overcome.

Population Growth and Distribution

One of the highest barriers is still the chain of poverty and population growth. In many regions rapid population growth continually adds to the burden of poverty. Current growth rates mean that every month more than 7 million people, the population of Bolivia (or the state of Georgia), are added to the world, and 90 percent of the growth is in developing countries.[28] As populations and expectations rise, so do other needs multiply, particularly for food, energy, and jobs.

Perhaps the most visible effect of economic, political, and environmental change has been the growth of cities in developing countries. By the year 2000, two-thirds of the world's urban dwellers will live in cities in the developing world.[29] In Latin America, it is estimated that some 40 percent of the population will be living in slums and squatter areas.[30] The rates of urban growth in the developing world have exceeded even rates of population growth. Mexico City, for example, quintupled its population between 1950 and 1980. The population of some African cities is expected to quadruple between 1980 and 2000.[31]

The relationship between population size and economic progress is neither simple nor straightforward. Historical evidence and cases can be mustered to demonstrate both the positive and negative effects of large populations.[32] Still, rapid growth of population has clear implications for the adequacy of national and natural resources and for the health of individuals, particularly women of childbearing age.

Reducing rapid population growth is a dual task. On the one hand, improved access to family planning techniques and services is essential to ensure the health of mothers and reduce infant death rates, as well as to expand the economic and social choices of families. On the other hand, much recent research has emphasized the importance of economic vitality itself as a motivation in the limitation of family size.[33] Employment, education, and social mobility all affect the choices families make about childbearing, whatever the techniques available.

The Debt Crisis

A second critical barrier facing development is the persistent debt crisis. The total debt of developing countries has reached some $1.2 trillion, $520 billion of which is owed to commercial lenders and the remainder to governments and international organizations.[34] The resource drain from developing to industrialized countries now totals some $60 billion annually,[35] a sum larger than the annual ODA transfer from donors to the developing world and a complete reversal from the 1970s. Although some countries have negotiated debt relief, the burden for many others remains crushing. In Mozambique, Somalia, and Sudan the value of annual export earnings is not enough to meet scheduled debt service payments.[36] The cumulative legitimate and illegitimate flight of private capital from developing countries to safer havens in the industrialized nations may equal total debt. As countries seek to reduce their debt, future private capital requirements must be kept in mind: adjustments will have to be made on a case-by-case basis so that economic conditions will improve reliably.

Security concerns, Yemen.
(Photograph by Christopher Warren.)

Human Rights Abuses

Human rights abuses continue throughout the world, even where political changes have begun. Censorship and other forms of human rights abuse are often closely associated with failures in development. Those who starve often starve in silence.[37] For countries with a free press and democracy, the risk of famine is reduced. By the same token, without freedom of inquiry and expression, knowledge, discovery, and innovation are imprisoned. Cooperation for development through science and technology cannot be effective in the long run unless it is also committed to the human rights of everyone.

War and Unrest

Violent ethnic or tribal conflicts and civil wars not only block development but destroy its achievements. New means of mediation and conflict resolution will be essential for clearing the barriers to development. Table 1 shows that there have been more than 100 violent conflicts between 1950 and 1990, and at least a score of comparable situations could be counted today. Indeed, centuries-old ethnic discrimination and conflict often intensify while democracy establishes its foundations.

Table 1. Violent Ethnic and Tribal Civil Conflicts, Civil Wars, and Cross-Border Conflicts, 1950–1990 (by location)

Region	Number
Latin America	13
Europe	5
Middle East	17
South Asia	10
Africa	30
Southeast Asia	37

Source: Ruth Leger-Sivar, *World Military and Social Expenditures, 1991*, World Priorities, Washington, DC, 1991, pp. 22–25.

Boom-and-Bust Cycle

Also impeding progress are the fluctuations of the global economic system. Economic slumps affect the willingness of donor countries to increase development outlays, of private enterprises to invest more, and of developed countries to import manufactured goods from developing nations. Simultaneously, they make the need for such increases greater by worsening the condition of developing countries. Periods of economic boom draw developing nations into the world system, only to be followed by busts in which they are bitterly marginalized again.

Many of the economic problems sketched above could be avoided with more prudent economic policies. Governments must implement market-oriented reforms and, as outlined later in this report, must foster pluralistic analysis of the options in policymaking.

Doubt and Skepticism

Within the technical fields of development itself, there continues to be a poor understanding of what works best in assistance and how to apply donor money most reliably and efficiently to leverage development progress. Indeed, there is debate, sometimes healthy but often debilitating, on what development "progress" really is. Given this poor understanding, development professionals face continuing conflicts over the objectives and policies when they recommend programs, sectors, and industries for support. Coordination among donors, and by performers of development projects, remains elusive; time and money may be wasted when coordination is poor.

Recognizing the uncertainties about both the effects of assistance and the goals of development, the public maintains considerable skepticism about public foreign assistance even as it gives generously to private charity. In the United States, this skepticism is expressed in three related criticisms: assistance is simply not effective; where it is effective, it takes jobs from U.S. workers; and the United States, with its own economic problems, cannot afford assistance. Whatever the merits of these views in specific cases, they are widely and strongly held; and their criticisms have been politically effective, especially when domestic needs appear to be in competition. That the American public translates its compassion into continued support for private philanthropy, but not for governmental foreign assistance, is attributable not only to a healthy concern for the public purse but also to a failure on the part of government to demonstrate the national interest associated with cooperation for development.

The Burden of the Past and the Challenge of Change

Of the greatest import, institutions remain locked into past conflicts and competitions. Military budgets drain huge resources not only from donor countries, but also from developing nations. In 1990, $880 billion was spent on armaments and training for war throughout the world, a total fifteen times the annual expenditure on official development assistance.[38] Although troop levels of industrialized countries have remained stable over the last three decades, military budgets have doubled.[39] In developing countries, troop totals have doubled and military spending has quintupled.[40] As a result, poor countries spend two or three times as much on the military as they receive in aid from donor nations.[41] In some countries, despite the persistence of disease, high mortality, poverty, and illiteracy, military budgets are many times larger than those for social needs (Table 2).

Meanwhile, bilateral aid agencies have often failed to recognize and respond to changing world opportunities and events. Innovations in technology and new scientific insights are poorly or belatedly linked to program priorities or planning. Too frequently, assistance clings to the pattern of past alliances, demands immediate and visible results to retain public support, and fails to plan for long-term needs or support vital underpinnings of research. Ultimately, the charge is made that too much is transformed into support for large governmental bureaucracies of both donors and recipients rather than for the individuals in the cities and villages for whom the programs were conceived.

Program balance has also been difficult to achieve in the midst of change. This has been obvious, for example, in the tension between the ongoing needs for agriculture and rural poverty compared with the soaring demands of cities, where the majority of people now live, though urban

Table 2. Military Spending as a Percentage of Health and Education Spending (1986)

Country	Percentage
Iraq	711
Syria	445
Iran	333
Pakistan	279
Peru	217
Saudi Arabia	155
South Korea	153
Jordan	150
China	146
Thailand	95
India	81
Cuba	79
Nigeria	63
Chile	59
Philippines	55
Guatemala	52
United Kingdom	45
United States	37

Source: The World Paper, February 1992, calculated from the *Human Development Report 1991*.

housing and other services lag badly behind population increase. Reconciling the ongoing needs with rapid change in development programs requires both more aggressive application of what we already know and greater attention to opportunities that might be realized through research.

Multilateral agencies have also had difficulty adapting to new needs and approaches. If they remain oriented toward large, centralized, public infrastructure investments in areas such as water and energy supply, multilateral agencies may be less suited to the new waves of development that emphasize markets, private property, incentives for individual enterprise, and the globalization of the marketplace. Multilateral agencies need to encourage recipient governments to promote open markets, political and economic decentralization, and individual and business initiative. Governments remain responsible for operating public infrastructure, and international financial institutions will retain their role in assisting governments with that task. In the future, skills and approaches will be needed within multilateral institutions to balance support for the roles of private commerce and public infrastructure in cooperation for development.

Underutilization of Science and Technology

Another problem—not really a barrier, and perhaps more properly seen as an opportunity—is that too little of the great power of modern science and technology has been directed at development. Mobilization of developed-country scientists to deal with problems found mainly in developing countries has not been very successful. The scientists and their employing institutions often lack the first-hand knowledge of conditions in developing countries that is required in order to formulate appropriate research agendas. Conversely, the cost of creating the infrastructure for developed-country scientists to do research within developing countries is often prohibitive, and long learning periods are required before such research is productive. The set of International Agricultural Research Centers is evidence that such an approach can be effective, given patience and commitment. The distributed network of scientists working in the Tropical Disease Research Program of the World Health Organization provides another successful, contrasting model.

The major Northern institutions concerned with development, science, and technology have a large unfinished agenda. They are not connecting enough with changing needs, people, and institutions elsewhere in the world. They urgently need help in recognizing this fact and in developing entirely new distribution systems to reach partners.

Meanwhile, as shown in Table 3, only about 4 percent of the world expenditure on research and development and about 14 percent of the world's supply of scientists and engineers are in developing countries, which contain about 80 percent of the world population.[42] There are enough successes in fields such as health and agriculture to suggest that much more could be profitably invested. For example, estimates of the economic returns on investments in international agricultural research indicate that these are much larger than for nearly every other type of investment.[43] There is a need to create understanding among leaders in developing countries of the benefits from such innovations. Better access to knowledge and information about available technological options is a requisite for the broader realization of the potential of science and technology serving long-term development.

Global Cooperative Development

The changes now under way allow the world to move away from merely preserving an armed truce, with its chronic, and sometimes acute, ideolog-

Table 3. R&D Scientists and Engineers and R&D Expenditure, 1990, by Region

	R&D Scientists and Engineers			R&D Expenditure		
Region, Area	Estimated Number	Percent of Total	Estimated Number per Million Population	Estimated Amount in Millions of U.S. $	Percent of Total	Percent of GNP
WORLD TOTAL	5,223,615		1,000	452,590		2.55
Industrialized Countries	4,463,800	85.5	3,695	434,265	96	2.92
Developing Countries	759,815	14.5	190	18,325	4	0.64
North America	930,720		3,360	193,720		3.16
Europe	1,091,000		2,210	104,960		2.21
USSR	1,694,430		5,890	55,710		5.66
Africa (excluding Arab states)	34,960		75	750		0.29
Asia (excluding Arab states)	1,190,360		400	88,530		2.08
Arab states	77,260		360	3,080		0.76
Latin America	162,930		365	2,360		0.40

Source: Table derived from the 1991 UNESCO Statistical Yearbook (UNESCO, Paris).

ical tensions, toward achieving the universal aspirations for peace, democracy, economic growth, and improvement in the quality of life. The changed conditions allow us to think and act as one extended global family. "Partnerships for global development" is the appropriate theme. Only when social and economic progress is widespread and durable will new democracies become deeply rooted and stable. And only then will newly free peoples be confident that the risks they have taken in turning toward the future will buy them the rewards of lasting liberty and prosperity.

WHY THE UNITED STATES?

In all of this change and opportunity, what are the interests of the United States in cooperation for development? A compelling answer to this question is essential for mobilizing sustained public support for the allocation of scarce public and private money for cooperation. Why should the United States not revert to isolationism?

Moral Interests

The readiness of Americans to respond to the needy and to meet adversity with an outpouring of generosity is an abiding strength of American culture. It has always been and it remains the foundation of public support for U.S. foreign assistance programs. Generosity and humanitarian concerns are a hallmark of American values with which people around the world identify.

At the same time, many deep and serious "development" problems confront American society itself. Applying the moral values that support development beyond U.S. borders provides an opportunity to reaffirm them within America itself. Global partnerships will lead to learning and action at home and abroad.

Economic Interests

Global prosperity is crucial to continued prosperity in the United States. In 1950, U.S. exports and imports accounted for under 5 percent of GNP; in 1990 they comprised 28 percent of GNP.[44] Indeed, between 1986 and 1990, U.S. merchandise exports accounted for 41 percent of the increase in GDP, and in 1990 alone they accounted for 88 percent of GDP growth.[45]

The U.S. Department of Commerce estimates that U.S. merchandise exports supported 7.2 million American jobs in 1990, an increase of 42 percent over 1986.[46] This level of job creation accounts for 25 percent of the growth in civilian jobs in the U.S. between 1986 and 1990.[47] Another reflection of the growing interdependence of the world economy is that approximately $2 trillion of assets in the United States are already foreign-owned (with the United Kingdom being the leading investor) and support 3.7 million American jobs.[48]

U.S. exports to developing countries exceeded $127 billion in 1990, much of it by small and medium-sized businesses.[49] It is estimated that a 5 percent decline in U.S. exports would cut real GNP growth in America by a fifth.[50] America's jobs depend on the health of its trade.

The markets that will buy and use what the United States can and would like to produce in the future—aerospace products, pharmaceuticals, software, for example—are markets of societies that are far more developed and wealthier than those in which most of the world's people participate today. Growth in the economies of America's trading partners and enhancement of the education and well-being of their peoples are critical to growth in U.S. exports and the possibility for balanced U.S. trade.

Foreign investment by U.S. firms already matters greatly to the U.S. economy and to American corporations. U.S. companies increased their foreign investments sixfold in the last three decades, from $56 billion in 1967 to $373 billion in 1989, of which $90 billion is in developing countries. Such investments benefit American companies not only in repatriated profits, but also with sources of goods and services that keep American companies competitive in the U.S. market. Fully 15 percent of all U.S. imports come from foreign affiliates of U.S. corporations.[51] The capacity of developing countries to become partners in such investment, through sound policies, financial strength, human resources, and infrastructure, will be of growing importance to the American private sector.

Just as each nation's economy has become more thoroughly entwined with the global whole, so U.S. economic interests hinge on policies that promote the viability of the world marketplace. Reinforcing the role of GATT[52] and facilitating the participation of developing countries in GATT are central to a stable trading environment from which U.S. exports can benefit. The globalization of markets means that the United States must formulate its policies toward all nations collectively rather than in isolated segments.

Ultimately, poverty elsewhere in the world hurts U.S. economic interests by breeding political and economic turbulence. This, in turn, disrupts financial and export markets upon which the United States depends. With the globalization of transportation, markets, and information, the industrialized countries are vulnerable to instability caused by social tensions, epidemics, extreme economic disparities, and large population movements virtually anywhere in the world. It is also important to note that social upheaval in the developing world slows the shrinking of military expenditures by the United States that might otherwise fuel U.S. economic growth.

Nevertheless, promoting liberal trade policies is not easy. Although the net aggregate gains of open trade far exceed the net losses, openness does create losers as well as winners. Ways must be developed to assist those (at home and abroad) who are temporarily disadvantaged by open trade to adjust to the structural changes so they can compete more effectively, continue to grow economically, and shift jobs from declining to growing industries.

Security Interests

Development is essential if four core U.S. security interests are to be realized. Security here is defined broadly.

Confidence in the Future

America's hopes for the worldwide advance of democracy depend on economic and social progress. The perception of such progress underlies the confidence of peoples that, however difficult the current change, they are now the masters of their own future. By contributing to that progress, cooperation for development helps maintain the momentum of the positive political and economic changes now taking place around the globe.

Peaceful Conflict Resolution

Continuation of the momentum of democracy is essential to the reduction of conflicts within and between nations. Democracies seldom declare war on one another. Although violence will no doubt continue to occur within and among nations, peaceful conflict resolution has become ever more essential in an era of weapons proliferation. Modern weapons mean that potential gains from violence can be offset in a few moments by the destruction and suffering that are the by-products of the violence. Development can contribute to peacekeeping by expanding prosperity and thus raising the penalties associated with extreme or prolonged violence. After all, those who have nothing have nothing to lose. New roles may emerge for the United Nations, and for regional groups, to resolve intranational conflicts.

Mexico City, Mexico
(Photograph by Christopher Warren.)

Prosperity

Large disparities in living standards among the world's people are not only economically destabilizing, they are also increasingly dangerous, both socially and politically. Tensions resulting from such disparities spread throughout nations and around the globe. They increase the pressure along fault lines of conflict, such as ethnic identification. The wealthy countries are unable to insulate themselves from the troubles of the poor, not even from their diseases. The spread of AIDS and the recent re-emergence of tuberculosis and cholera may be harbingers of future epidemics spread in all directions by the deadly mix of poverty and mobility. To reduce global risks, wider prosperity is essential, along with better-designed national and international policies and regulations in fields such as health and environment.

Global Environmental Quality

The winds and waters of the world flow freely across borders. The people of the United States cannot be assured environmental security unless the rich and poor of the world cooperatively address issues of population growth, industrial practice, and land and energy use.

Scientific Interests

Cooperation is needed not only for economic and social development, but for the progress of science itself. Scientific benefits of cooperation always flow in both directions. Many developing countries have brilliant scientists and engineers. In the past, almost all U.S. research was carried out using solely domestic resources. Today and tomorrow, growing numbers of problems, ranging from computer development and climate change to AIDS and cancer, require research partnerships and data sharing among countries. The United States is advantageously positioned to contribute to such partnerships and to strengthen the institutions in the developing world that can become effective partners in scientific research that, in turn, will be of importance to the American people.

WHY NOW?

Now is a rare moment of historic opportunity to promote peace, liberty, and global prosperity through cooperation. It is a time for creativity comparable to the period immediately after World War II, and the chances of success are even greater. This is also the time to mobilize science and technology to speed change toward these goals. The deed is not yet fully done. The foundations of openness and freedom that have been laid in recent years in many corners of the world must be reinforced if they are to withstand the setbacks, disagreements, and reversals that surely lie ahead.

Cooperation for international development must rise to this challenge. Cooperation need no longer be about *buying* allies. The opportunity now is to transform development cooperation into *investing* in the world's prosperity. Tangible, visible, firm commitments to cooperation for development are essential if we are to ride the tide of individual, economic, and political freedom now rising worldwide. The time for commitment is now. To wait is to risk the gains so recently won.

UNIQUE ASSETS OF THE UNITED STATES

All nations of the world can and should step forward to accept the new challenge. To the opportunities and problems ahead, each nation brings special strengths. For its part, the United States has two preeminent assets as a development partner.

Leadership

Around the world, America is a symbol of freedom and progress. The Statue of Liberty rising over Tienanmen Square in 1989 exemplified the hope that America represents to those seeking freedom and democracy. The commitment of American leadership to cooperation for development sends the strongest possible signal to all nations that the United States stands shoulder-to-shoulder with those who accept the risks of economic and political change. It reminds everyone of the urgent need to take seriously a responsibility toward the world's poorest peoples. When the United States moves ahead to lead a coalition, the world gains. And when the United States hesitates, the potential for cooperation drains away.

America is also an example, if still imperfect, of a free and open market and the many benefits that it can bring. The U.S. private sector is the largest, most diverse, dynamic, and independent in the world. It offers models of all kinds of enterprises and innumerable partners. But the American private sector must become more committed to bringing its experience and insights to worldwide development, more dedicated to open and fair markets.

Beyond political symbols and entrepreneurial drive, America has 200 years of experience in building and, with the tragic exception of the Civil War, peacefully adapting policies and institutions within a decentralized democracy. The mistakes as well as the successes can be offered to developing-country partners to ensure that hard lessons need not be painfully learned twice, and that "what works" can be observed and applied. U.S. institutions are uniquely transparent to outsiders. By law and custom, most of the information, regulations, and processes of public departments and agencies are open to public study, scrutiny, and inspection. America's institutions of higher learning are also uniquely varied and accessible.

Finally, American resources available for development partnerships are far greater than shown by the measures of public budgets or private trade. The American tradition of voluntarism has meant that many individuals and organizations share knowledge and services for cooperation in development. No other nation has a cadre of comparable size and skills to participate in technical cooperation.

Strength in Science and Technology

The only way to solve the problems of societies and meet their multiplying needs is with enhanced contributions of science and technology to development and by extended cooperation between the "science-rich" and the "science-poor." The depth and breadth of U.S. expertise in science and technology is unrivaled. U.S. experience with application of technologies ranges from automation of financial systems to vaccination of inner-city children.

REDEDICATION

In sum, three realities define the case for reformed and reinvigorated partnerships for international development.

First, the Cold War has ended, and with it the distortion of international relations and economic cooperation that it brought. The 1990s offers the opportunity to shift assistance from the cause of political and military alliances toward the goals of global development.

Second, a massive restructuring of the world economy is under way. People and their leaders everywhere wish to build market-friendly models of economic development in place of state-led models. In literally scores of countries in Asia, Africa, and Latin America as well as Eastern and Central Europe, this change creates enormous opportunities to raise economic efficiency, enlarge participation in development, and strengthen freedom. It ensures that foreign technical and financial assistance will be more effective in the future. More importantly, private trade, investment, and technical cooperation by Americans with developing countries will have greater scope and impact.

But the restructuring of the world economy is not just a matter of what is happening in developing countries or the former Soviet Union. The economy of the United States is undergoing major changes, as well as adjusting to new forces in the world market that have abruptly made large portions of the economies of many nations obsolete.

The third reality is that terrible gaps in human progress persist, both within and among countries. Desperate needs in health, food, education, and other areas must be met.

Understanding these three realities is the path to resolving the puzzle of "why cooperate?" For the people of the United States must answer the inescapably consequential question of their future engagement with global development. The answer to this question is that the realities demand a firm rededication to the historic principles of U.S. cooperation.

The Task Force recognizes that rededication will be meaningful only if it transcends rhetoric. The first step toward making rhetoric an operational plan is determining the content of cooperation. Then there can be an *agenda of actions*, the "what to do" of highest priority for the next decade. That is the subject of the next chapter.

Vigorous rededication also requires *institutional reform*—the "how to proceed." Major organizational and legislative changes will be needed throughout the United States if partnerships for international development are to respond to the changing circumstances and the emerging opportunities. Those changes are the subject of the fourth chapter. Chapter 5 offers some final thoughts on what the past has meant and what the future might bring.

3
WHAT: THE CONTENT OF COOPERATION FOR DEVELOPMENT

As the reasons for cooperation for development — the "why"— have evolved, so must the substance, the "what." The problems faced in development have altered and diversified. The trajectories of development involving industrialization, urbanization, and demographic and other changes, as well as the different rates at which countries travel along them, are better understood. Furthermore, countries recognize more clearly that they travel on trajectories that bring collective risks, such as global warming and loss of biological diversity. These must be faced by all nations together as part of the development process.

To remain relevant, therefore, the agenda of cooperation for development must adapt. The United States, as the largest donor and the partner with the most deeply established practices, has the most difficult adjustments to make to respond to the new needs and opportunities. Its success will be correspondingly significant.

Here the Task Force first summarizes the larger patterns within which development takes place. The main focus of the chapter is on the *principles and criteria* that should govern the choice of programs for development for the next decades. A more flexible system for cooperation is mandatory in this period of transition and increasing diversity of needs.

Unfortunately, the present state of affairs is unsatisfactory. The final section of this chapter offers a new approach, using several concrete examples to illustrate how the recommended principles and criteria should be applied.

LEVERS ON DEVELOPMENT

There are many potential levers on development, some public and many private. For the U.S. Government, these begin with programs to ensure security and peace that set the framework within which development can occur.

The U.S. Government also has many other "international" policies and programs that have consequences for international development. These relate to trade, environment, energy, drugs, migration, foreign students, and intellectual property rights, to mention but a few. At least as important are "domestic" policies and programs that affect American savings, investment, and economic growth, and inevitably the world economy.

Of course, macroeconomic performance in the industrialized countries affects development of the rest of the world. The Task Force fully recognizes that overarching macroeconomic forces determine how smoothly development proceeds, how resources become available for it, and how effective U.S. bilateral efforts can be.

This report focuses on U.S. and multilateral programs and organizations whose primary and direct purpose is international development. Nevertheless, the effects of many other "levers" on development are so great that they must be factored into the national and governmental processes of decision making for development.

PRINCIPLES OF BALANCED INSTITUTIONAL DEVELOPMENT

Even if much greater financial resources and political attention were devoted to development, programs would not function wisely and well in the absence of sound principles on which to base the substance of partnerships. Experience with cooperation in development, whatever the subject of programs, teaches that long-term benefits emerge most strongly when the choice and design of programs are guided by well-understood principles.

The most fundamental principle of cooperation for development is to foster the balanced development of the public, private, and independent sectors, pluralism within these sectors, and creative interaction among the different ways of thinking that underlie the sectors and the institutions in them. Cooperation of this kind can be achieved only by balanced participation of the different sectors in the donor countries as well. Cooperation must conjointly build sound public administration, an enterprise culture, and a lively, critical independent sector.

Private Sector

Governments alone cannot generate economic and social progress. The engine for development is the enterprise culture, which is concerned above all with results. This culture protects the autonomy and freedom of the individual, promotes innovation, and ensures that private investment generating social benefit is rewarded. It facilitates the formation of flexible networks to accomplish tasks on all scales as efficiently as markets allow. The private sector is the source of most wealth for investment, whether indigenous or foreign. The "bottom line" is the principle that creates a vigorous private sector (see Box 2).

Propulsion for the enterprise culture comes from science and especially technology. They are important enabling tools in a responsive market-

Box 2. The Enterprise Funds

In seeking to encourage the emergence of an entrepreneurial culture, free markets, and private competition in the newly democratic nations of Eastern and Central Europe, the United States, through the Agency for International Development, has created four "Enterprise Funds" in the region, one each for Hungary, Poland, the Czech and Slovak Republics, and Bulgaria. The objective of the Enterprise Funds is to promote policies and practices conducive to the development of the private sector through the provision of loans or equity investments in small and medium-sized businesses. Although initially capitalized with U.S. Government funds, they function as private investment entities, completely privately managed according to standard business practices and governed by Boards of Directors made up largely of private-sector executives and bankers from the United States and partner countries. They are also empowered to function as financial wholesalers, soliciting monies from other investors and thereby creating a larger pool of capital. Profits generated by Enterprise Fund investments or loans are retained and reinvested in new projects. Among the Enterprise Fund investments have been projects in housing, agribusiness, and business technology.

place. Technological innovation enhances and speeds the flow of information, which lowers the costs and guides the energy of entrepreneurs. Scientific research and technological development underpin innovation, facilitating the creation of new services and products in response to individual choice and increased freedom. In turn, the entrepreneurial culture serves science and technology by speeding the dissemination of innovation through the global marketplace.

Public Sector

The public sector must provide and strengthen the organizational and administrative means to build and maintain democracy, honest political and economic processes, representative government, and the rule of law. The public sector also has special responsibility for fairness in markets, sound regulation to balance privilege and obligation, prudent use and protection of the resource base and the environment, and patronage of basic research and other public goods.

Stable public or quasi-public entities are needed to address and administer numerous large and complex social functions. Sophisticated government policies and competent governance are required for military and social security, environmental protection, health care, transport, communications, and other infrastructures. Some of these functions, for example family planning, can be promoted through private initiative, involving both nonprofit and for-profit groups. Governments must have the ability to gather and analyze data, frame options, evaluate results, and articulate choices to the electorate as regards national priorities. Sound public administration relies on carefully codified and universally respected processes. A modern public sector cannot operate without massive applications of science and technology—for example, in handling the flows of information needed for overseeing banking, health, and environmental goals.

Independent Sector

It is vital to maintain social organizations that check inequalities of wealth, status, power, and knowledge, that protect human rights, that promote voluntary cooperation among the individuals and groups, and that provide for the tailoring of new ideas and products to particular cultural contexts.

Local groups and initiatives are the sources of creativity and persistence in pushing for humane innovation and responsible government. Universities, churches, many nongovernmental organizations, and the "independent sector" more generally flourish in a society that embraces critical debate as an ingredient for growth that is both sustainable and equitable. This sector promotes political pluralism, freedom of religion and the press, the rights of minorities, and direct expression and consent of the governed.

It is also the independent sector that is most active in meeting the needs of those marginal members of society who will never be absorbed by either the enterprise engines or routine administrative practices. It is often most effective at delivering services to the most needy.

Science and technology are indispensable to a healthy independent sector: they provide the expertise to balance that retained by otherwise more powerful interests, and they foster a culture that does not take received wisdom for granted.

A Balanced Approach

A stable democratic society arises from the healthy interaction of diverse ways of thinking in a humane context. For much of its history, development assistance has involved an emphasis on only one sector or approach, with predictable shortcomings in the results. Cooperation for development must encourage balanced evolution in societies of the knowledge, organizations, and decision-making processes utilized in each of these sectors. In some societies, partnerships for development must explicitly take on the task of building an enterprise culture, or an independent sector, or sound public administration—and, in all these areas, science and technology play essential roles.

CRITICAL ROLES OF SCIENCE AND TECHNOLOGY

Science and technology are critical to a new approach to cooperation for development. Ultimately, the key components for building prosperity are knowledge, skills, and liberty. Science and technology

- Undergird the research that creates needed knowledge
- Help build the education and training systems that advance skills
- Thrive with the freedoms of inquiry, communication, and association that ensure, and are ensured by, democracy and liberty

All major developmental goals—rapid economic growth and industrialization, environmental protection, modern telecommunications, improved

Box 3. Halving World Hunger by the Year 2000

A major initiative is under way to end half the world's hunger before the year 2000 by combining the products of recent scientific research with new technologies of implementation, consensual approaches, and pluralistic institutions. Simultaneously, longer-term S&T innovation is essential to sustain and extend the accomplishments that are possible in the near term.

Although numbers are imprecise, some one billion people live in households too poor to obtain the food they need in order to work; a half a billion in households too poor to obtain the food they need in order to move around. One child in six is born underweight, and one in three is underweight by age five. Hundreds of millions of people suffer anemia, goiter, and impaired sight from diets with too little iron, iodine, or vitamin A. Yet there is a consensus that, by linking what we now know with what expanded S&T research can tell us, we can reduce by half the toll of hunger within a decade.

This is the conclusion of groups and agencies concerned with hunger, including the World Food Council, the Task Force on Child Survival, heads of state at the World Summit for Children, and the authors of the major nongovernmental initiative developed at Bellagio, Italy, in November 1989. Specifically, the Bellagio Declaration proposed four achievable goals for the 1990s: (1) to eliminate deaths from famine; (2) to end hunger in half of the poorest households; (3) to cut malnutrition in half for mothers and small children; and (4) to eliminate iodine and vitamin A deficiencies. In the three years since the Declaration, there is much progress to report.

Famine. Leading the agenda is the potential for virtual elimination of deaths due to famine among the 15–35 million people annually at risk, through implementing existing early warning and famine prevention systems and, most important, through continuing efforts to provide safe passage of food in zones of armed conflict.

Nutrition. Equally capable of eradication are two of the three major nutritional diseases. By iodizing salt or injecting iodized oil, most of the 190 million cases of goiter could be eliminated by the end of the century. A capsule given twice a year to the 280 million children at risk of vitamin A deficiency could virtually eliminate the disease in the crucial ages between 1 and 4 years, not only preventing blindness but dramatically increasing child survival. A coordinated international effort is under way to eliminate vitamin A and iodine deficiencies and reduce iron deficiency. It will combine the hitherto competing approaches of supplementation (pills and injections), fortification (additives to food), and dietary change (use of mineral and vitamin-rich foods) to address the "hidden hunger" of micronutrient deficiencies.

It is also possible to halve malnutrition among women and children. Rapid progress has been made in immunizing infants and providing simple, home-based treatment of diarrhea. Breastfeeding of infants is continuing or even increasing in many developing countries. Innovative programs in Africa and Asia combine the monitoring of growth by weighing the child with supplemental feeding as needed. These activities can be combined with efforts to ease the burden on already overworked mothers and to reduce the nutritional anemia found in half of all women of reproductive age.

Poverty. Most hunger is rooted in poverty, but the hunger of at least half of the poorest households can be ended. Extensive experience with food subsidies, coupons, ration shops, and feeding programs demonstrates that careful targeting and effective application of such measures could reduce much urban food

Box 3. *(continued)*

poverty at relatively low cost. In rural areas, providing wage and food income in return for labor to construct needed agricultural and environmental improvements reduces food poverty immediately while simultaneously increasing long-term agricultural productivity and income. Other programs provide self-sustaining sources of credit, especially to women, to start small businesses or to produce local products and services.

Subsistence Farming. Food-poor households that raise their own food have to cope with the deterioration of their natural resources, the loss of crucial access to common resources, and restriction to all but the most ecologically marginal land. There are important opportunities for redistribution to smallholders of land that is little used, and a variety of low-cost techniques have demonstrated ability to sustain productivity, provide fuelwood, limit soil erosion, and increase food and income.

Costs. A systematic assault on the hunger problem will inevitably require coordinated action in new partnerships, as well as new resources and food aid, linking the rich to the hungry and poor. One estimate of the cost of a realistic program to combat hunger in the 1990s is about $8 billion a year, half in new resources (with some from the United States and most from a pool of donors including Japan, the EC, and international institutions). This is the equivalent of about $7.25 for each citizen of the affluent world. Even more important than new resources is the need for renewed social support and political will and the creative employment of local institutions and underutilized resources in order to encourage incentives for independent economic growth. A careful plan would have to be developed on both the substance of action and the mobilization of resources, spanning investments in food aid as well as in research on such topics as fast-growing trees and other crops, production on arid lands, and biotechnology for agriculture in the 21st century.

Private Volunteer Organizations. Private voluntary organizations are particularly important in reaching the hungry and poor, and the best of them are hungry people acting in their own behalf. The most promising approaches empower people to assess their own condition and to act to improve it, provide short-term hunger relief while addressing deeply rooted causes, and demonstrate sustainability over the long term. The last decade has also witnessed a slow emergence of new public voices for the hungry and impoverished.

What Can Be Done Now. There is now in place new scientific knowledge to provide early warning of famines, to break the nexus between hunger and disease in children, and to increase food production in sustainable ways. There are also new technologies of implementation: ways to target assistance or provide credit to the very poor, ways to immunize a whole generation, and ways to provide mothers with the tools to monitor their children's growth and health. There are new approaches that combine markets with safety nets and link the grassroots with the summit by new networks of institutions.

In the Long Term. While near-term options are many, as population grows, long-term food sustainability will require expanded investment in research on agricultural systems and technologies. This is especially important if expanded production and growing populations are not to result in a further deterioration in fragile ecologies. Similarly, long-term investments must be made in instituting the types of stable incentives for farmers that will encourage production and discourage environmentally dangerous practices.

health, better farming, responsible population management, new techniques for housing and biotechnology—depend to a large degree on the ability of countries to absorb and use science and technology. Box 3 illustrates how, through science and technology, the known and the new can be harnessed to address urgent development issues, in this case the goal of halving hunger by the year 2000.

The countries now succeeding in a world that is increasingly science-based are those that have long invested in scientific education and technology training and learned how to use the results. Cooperation can assist less-developed countries to make the investments in science and technology that will provide them with the human and technical resources to fuel sustainable development.

CRITERIA FOR PROGRAMS

Traditionally, cooperation for development has been approached by specific "sectors," with agriculture, energy, and transportation in the lead. Fresh approaches must emphasize fundamental social conditions and the forward-looking policies that are conducive to sustained prosperity and that in turn can be enhanced by it. In any given case, a candidate program for a partnership in development must be examined in terms of the urgency of the problem and the intrinsic merit of the ideas for its solution.

More broadly, there are four criteria for selection, design, and conduct of programs: the policy environment in the recipient country in which the opportunities exist; ecological and social sustainability; building the capacity in a country to solve its future problems; and cultural sensitivity and mutual respect between the partners, creating common objectives and shared responsibilities. Fulfilling these criteria significantly raises the chance that the partnership will contribute to sustained development. Cooperation, in turn, should be designed whenever possible to promote the objectives that these criteria embody. Each of the criteria requires explanation.

Favorable Policy Environment

The policy environment spans all sectors of society. Among the policies most important for development initiatives are fiscal and monetary policies that in combination promote noninflationary, sustained economic growth; trade policies that favor competitive excellence in domestic industry; policies for efficient use of resources; and policies that protect property rights. Stable *macro*-economic policy environments must be in place to create the *micro*-economic conditions that encourage individual entrepreneurs. Entrepreneurship is sufficiently risky by itself—it should not be burdened with the added risk of unpredictable and rapidly changing macroeconomic policies and unnecessary regulatory intervention. Box 4 describes ways in which development assistance has worked to create strong economic policy capability in developing countries.

Open markets and trading are essential. This is as important in the policy corridors of industrialized development partners as in those of developing countries. Yet, unfortunately, by the end of 1990, GATT members had in place over 250 arrangements to impede the flow of developing-country products into their markets. The United States has encouraged a more open approach to its markets than many of its industrialized counterparts. As can be seen in Table 4, America absorbs the greatest share of developing-country exports. The United States can make a major contribution to development by encouraging other industrialized nations to join in truly global approaches to open markets and trade, with particular attention to the goods of the developing world. Protectionism is, quite simply, everywhere an enemy of long-range development.

But the "policy environment" goes beyond economics: peace, political freedom, and pluralism are preeminent considerations. Without these, development is unlikely to contribute to sustainable economic or social progress for individuals. Indeed, as many countries achieve economic progress, their peoples demand greater freedom and liberty. Without such freedom, the progress itself may be threatened. Among the most important policies are those for protection of freedom of expression, education for a well-informed public, and assurance of free and fair elections. Policies must promote equity among national groups and between the sexes and due process for all. These concepts should be encouraged by the process and content of official development assistance.

Ecological and Social Sustainability

Sustainability means that choices made today about economic and social development expand, rather than restrict, the choices available to future generations. Sustainability is, of course, a dynamic criterion. Technological innovations, as well as political, economic, and social change, can dramatically affect assessments of sustainability. Population growth is a strongly determining factor. Sustainability is an indispensable criterion for development, and its achievement is feasible.

Using the experience of forest protection, Box 5 illustrates the types

Box 4. Economic Policy: Building Capability and Institutions

Development assistance through bilateral and multilateral channels during the 1950s, 1960s, and 1970s focused primarily on investment projects in infrastructure, agriculture, industry and, to a lesser extent, the social sectors. By the 1980s it became clear that many of the projects had failed to yield the anticipated rates of return. One reason was that investments were made in a highly distorted policy environment. Investments to improve agriculture through irrigation, credit, infrastructure, and technical assistance often failed when government policies favored urban consumers at the expense of rural producers, primarily by keeping prices for food artificially low. Similarly, a distorted structure of prices for consumer goods and intermediate inputs, mostly a result of inadequate trade and credit policy interventions, reduced the necessary incentives for efficient private investors and subsidized inefficient investments.

A few developing countries, notably South Korea, Singapore, and Taiwan, introduced policy reforms early to correct harmful distortions. By the end of the 1980s most developing countries, prodded in many cases by multilateral institutions, accepted the need to introduce policy reforms to seek a better balance between market forces and state intervention. This began to produce trade and financial liberalization, privatization, and improvements in regulatory practices. Much more remains to be done.

These developments highlighted the importance of the government's capacity to formulate and execute policy reforms, of academic and independent institutions conducting policy-oriented research, and of private sector capabilities to assess the impact of policy reforms on the performance of manufacturing and service firms. Fellowship programs, short-term training, research grants and contracts, institutional support, and small grants for events, publications, and travel have been the main mechanisms for building capacity for economic policy.

Fellowships. The Colombo Plan in South and Southeast Asia and the Ford Foundation in Latin America had major fellowship programs during the 1950s, 1960s, and 1970s that supported graduate students in the social sciences, primarily in U.S. and European universities. The Ford Foundation is credited with helping to train the first generation of professional economic policymakers in several Latin American countries. Similarly, the Fulbright program administered by the U.S. Government has enabled students from all over the world to do their graduate work in the United States in a variety of fields, with a good proportion choosing the social sciences. More recently, the Japanese government has provided resources to the World Bank and the Inter-American Development Bank for a graduate fellowship program in economics, international relations, and related disciplines; this program supports hundreds of students from developing countries every year, as they pursue their studies in the United States, Japan, and elsewhere. As a result, a large number of the economic policymaking elite in many developing countries have been trained at leading universities in the United States and Europe.

Box 4. *(continued)*

Short-Term Training Programs. These programs have been and are offered by a variety of private, bilateral, and multilateral institutions. One example is the highly regarded 8-week training course for staff members of central banks offered jointly by the U.S. Treasury Department and the World Bank every year. The IMF and the World Bank also offer short courses in a variety of topics, ranging from stabilization policies to project design, privatization, tax reform, and economic policy management. In addition, several private institutions offer training programs to developing country nationals, mostly under contract with bilateral agencies such as AID.

Research grants and contracts. One of the most important factors in improving economic policymaking capabilities in developing countries, and particularly in Latin America, during the last two decades, has been the creation of university-based and independent policy-oriented research centers. In some cases with government and private sector support, and mostly with external funding from bilateral agencies, multilateral institutions, and private foundations, these centers have conducted empirical studies, developed policy options, organized debates and seminars, and published books, reports, magazines and journals.

U.S. foundations, especially MacArthur, Carnegie, Ford, Rockefeller, and Pew, have been particularly active in this area, mainly in sector-specific economic policy research in agriculture and health. The Canadian International Development Research Center (IDRC), the Swedish Agency for Research Cooperation with Developing Countries (SAREC), the Netherlands University Foundation (NUFFIC), and the Sasakawa Peace Foundation are also among the many private and public financing institutions that provide research grants in the social sciences. In addition, multilateral institutions, including the World Bank, the Inter-American Development Bank, and the European Community, and bilateral agencies such as AID, the Canadian International Development Agency (CIDA), and the British Overseas Development Institute (ODI), give contracts and in some cases grants to economic policy research centers in developing countries. A recent and particularly effective way of supporting policy-oriented research has been the network of economic policy research centers, established by the Inter-American Development Bank, which transforms the results of research projects into policy options and proposals for implementation.

Latin America is the developing region where independent research centers have flourished and wield considerable influence. CIEPLAN and CLEPI in Chile, FUNDESARROLLO in Colombia, Fundación Mediterráneo and the Instituto di Tella in Argentina, GRADE and the Instituto Libertad y Democracia in Peru, and IESA in Venezuela are examples of strong policy-oriented institutions. In contrast, in many countries of East Asia, government policy study centers, which have enjoyed considerable autonomy, have been extremely active in conducting studies and exploring policy options. For ex-

Box 4. *(continued)*

ample, the Korean Development Institute has been for many years a center of excellence in trade, finance, and technology policy research. In many other regions, the establishment of independent policy research institutions is a relatively new phenomenon.

Institutional Support. The provision of external support to educational, research, and training institutions in developing countries to cover general expenditures, as well as program costs, was common in the 1960s. In some cases this took the form of a donation to establish an endowment, which subsequently generated resources to cover recurrent expenditures. During the 1970s most foundations and development assistance agencies switched to providing support for specific projects and programs. From the perspective of the donors, this allowed better monitoring and review, but the mechanism left recipient institutions without support for general expenditures. In turn, this required recipients to prepare and negotiate many small project proposals continuously, which often diverted their efforts from research and studies.

As the number of academic, government, and independent centers increased during the 1970s and 1980s, institutional support became rare and competition for external support intensified, particularly in Latin America and some East Asian countries, with the consequent fragmentation of funding and a reduction in the average size of grants. Some funding agencies have begun to reexamine this situation in the 1990s, and there is renewed interest in exploring ways of providing institutional support, particularly in view of the new emphasis on dissemination and the utilization of research results. These generally involve mass media activities, seminars, and workshops that are considered "overhead." Efforts to reduce administrative costs in funding agencies, often by increasing the average size of grants, have also contributed to this reexamination of the importance of institutional support. The African Economics Consortium, sponsored by the Rockefeller Foundation and the IDRC, is an example of the return to institution building, in this case reinforced by expanded networking within Africa itself.

Small Grants. Another form of support for capacity building for economic policy has been the provision of small grants to finance the incremental costs of publications, seminars, workshops, and other events. This has been the preferred approach of institutions such as the Friedrich Ebert, Friedrich Naumann, and Konrad Adenauer Foundations in Germany, as well as of small international foundations and of private corporations in developing countries. This has allowed research centers to obtain resources for specific dissemination activities, although the need to secure these funds can lead to the dispersion of efforts and generate inefficiencies.

In general, building the institutions required to examine the evidence for alternative economic policies objectively will be crucial to the success of market-friendly economic reform.

Table 4. Relative Share of Developing-Country Exports to Industrialized Countries, 1990

Country	Percent Share
United States	26
Japan	23
Germany	12
France	8
Italy	6
United Kingdom	6
All other OECD	19

Source: *Handbook of International Trade and Development Statistics*, UNCTAD, Geneva, 1990, Part A.

of partnerships that ought to be a priority in efforts to ensure that development programs and ecological sustainability are closely linked.

Building Capacity To Solve Future Problems

An essential aim of cooperation in development must be to enable partners to make, and act on, their own choices. Cooperation in development, whether concerned with forests, farms, malaria, or manufacturing, must increase and diffuse the local pool of general skills and, even more, strengthen the capacity of local institutions that store, add to, and share knowledge. But capacity building is not just about schools, colleges, and universities. It is about the education, access to information and communication, social learning, and learning-by-doing associated with every aspect of development; this includes, for instance, how to conduct an election and how to run a business at any scale and with any level of needed technology.

In sum, capacity building has five main aspects: the building of individual and group competence; the generation of relevant new knowledge; the diffusion of knowledge to potential users and the refinement of this knowledge through application; the building of the institutional infrastructure to support education, research, and diffusion of knowledge; and the enhancement of the capacity of public and private organizations to reach sound decisions.

Human resource development is at the heart of the realization of individual potential, for women and men, girls and boys. The emphasis must be on the ability of nations and markets to provide individuals with

Box 5. Protecting Forests

Forests, both in the tropics and in industrialized nations, are a critical element of global ecology. They are the lungs of the earth, absorbing a large fraction of the atmosphere's carbon dioxide and releasing oxygen. Tropical forests alone contain 60 percent or more of the world's plant and animal species. As well as performing environmental services, forests provide jobs and valuable economic products. Their continued viability is essential to sustainable development.

Yet forests are disappearing rapidly in many regions. Though few nations have taken reliable inventories of their forests, estimates are that just 1.5 billion hectares of undisturbed primary forest remain of the 6.2 billion that existed before settled agriculture began. During the 1980s, deforestation may have claimed as much as 7 million hectares per year of irreplaceable tropical forest. Destruction and degradation of forests is also an urgent issue in industrialized countries.

Deforestation, particularly in tropical areas, is fundamentally a result of failures to "value" forests fully and correctly and then to allocate returns realized on their value to forest management and sustainability. A concerted effort to ensure the sustainable management of forest resources requires new vehicles for global cooperation. It requires participation by scientists from both the developed and developing nations, and by the owners of tropical forests, predominantly governments, along with the harvesters and purchasers of forest products, predominantly the private sector.

Research is needed to define better the environmental services provided by forests, to estimate sustainable yields of forest products, and to understand the social aspects of forest management. In many regions significant forest management capacity must be built. The emphasis should be on participation by indigenous scientists and forest managers in developing countries to ensure that countries are themselves able to continue to monitor and manage their forest assets.

An example of cross-national forest management is the Sustainable

the integrated complex of health, education, and employment through which they contribute to national progress. Box 6 and Table 5 illustrate the importance of the U.S. university system in assisting developing countries on the path to these goals.

Partnerships as the Premise

The fourth and last fundamental criterion for development cooperation is the general notion of a partnership. The concepts of "donor" and "recipient" are outdated and must be laid to rest. True partnerships must be forged between countries, partnerships in which the expectations of the partners are clear, in which each has something to gain and each has a clear responsibility, with accountability, for progress toward goals in the program.

Box 5. *(continued)*

Management of Tropical Evergreen Forests Project in Asia, developed by the Harvard Institute for International Development (HIID). Using a standardized protocol, the project is carrying out an interdisciplinary inventory of forests in Sri Lanka, Indonesia, Malaysia, India, Thailand, and Brunei. These inventories, together with local training, will create the base for valuing the products of the forests and for managing their use so that the forests are preserved and revenues applied to maintain productivity of the forest resources. About half the initial costs of the forest assessments are paid by local sources.

The HIID project is designed to become a self-sustaining indigenous endeavor, as the wise exploitation of the forests generates funds needed to conduct continuing research and monitoring as well as provide employment and income. The project directly addresses shortcomings in the policy environment, including the perverse economic policies and institutional failures that create incentives for destructive harvesting and windfall profits.

Research is needed in many other aspects of forest maintenance as well. For example, pests are a major source of forest loss. Few environmentally friendly mechanisms are available to contain this threat to global forests. Extensive collaborative research between scientists of developing and industrialized nations is needed to devise pest control technologies that will benefit both rich and poor countries.

The relationship between forest restoration and other environmental problems represents another example of the type of collaborative S&T-based research that is essential to ecological sustainability in development. Forestry, drought, and desertification are closely linked in many countries, both developing and industrialized, and must be studied and attacked in tandem. The collaborative efforts of the African Academy of Sciences and the U.S. National Science Foundation in this area are an example of development cooperation whose results will benefit all nations engaged in the effort.

Source of data: *World Resources, 1992–1993*, Oxford University Press, New York, 1992.

Programs that represent such partnerships will not be ephemeral. Nor will they necessarily be simple. They will benefit from planning, careful attention to details in the field, the creation and mobilization of experience and expertise from both sides of the partnership, and steady management to resolve the inevitable conflicts and face up to failures revealed by constructive evaluations.

PARTNERSHIPS OF INTERESTS, EXPERTISE, AND MANAGEMENT

Balanced institutional participation, full use of science and technology, and perceptive application of these four criteria are the beginning, not the end, of the changes needed in cooperation for development. A key problem is

Box 6. U.S. Higher Education and Development

America has become a university for the world. For the one million students who now travel abroad to pursue their higher education, the United States is by far the preferred destination, with over 400,000 foreign students in 1991. About 60 percent of all Asian and Latin American students who study abroad come to the United States, about 25 percent of all students from the Middle East, and some 15 percent of all students from Africa. About three-fourths of foreign students in the United States come from developing or newly industrializing countries. Since the mid-1970s the number of foreign students in the United States has more than doubled. Two-thirds of the increase has come from Asia, while the number of students from Africa in 1990/91 was actually lower than in any year since 1975.

Most foreign students come for technical training. Typically, over 40 percent of foreign students choose to study engineering and science (including mathematics and the physical, life, and health sciences), and some 20 percent study business and management. Foreign students are about equally divided between undergraduate and graduate programs. Seventy-three percent of funds for foreign students is from non-U.S. sources. Of this, almost 90 percent comes from the individual and family, and only 10 percent from government. Of the funds from U.S. sources, most come from U.S. colleges and universities themselves. The U.S. Government is the primary source of funds for fewer than 2 percent of all foreign students.

Foreign students are found in all regions and in all kinds of institutions. The critical, independent spirit of U.S. universities, the high value placed on freedom of speech, and the social mobility U.S. universities embody provide "hands-on" education in democracy and pluralism. At the same time, students acquire whatever specific knowledge and skills they need.

Students from particular universities and departments have often formed tight-knit and influential groups later in their careers in their home countries. This is true, for example, of economics graduates of the University of Chicago in Latin America. In some fields more than half of all foreign students remain in the United States to pursue their careers, benefiting the United States but draining talent from their homelands. However, these individuals often build international bridges in commerce and culture even when they remain in the United States. Moreover, some foreign students who have made careers in the United States have later returned to their countries of origin and become major investors in economic growth. This "reverse brain drain" has been quite striking recently in Korea and Taiwan.

It is both futile and wrong to try to prevent the flow of human resources. However, much more can be done to use higher education in the United States to build capacity in developing countries. It is especially important to note that very few people from many of the poorest nations come to study. Multilateral institutions, bilateral institutions, and U.S. universities themselves could

Box 6. *(continued)*

substantially enhance their efforts to create educational opportunities for students from the poorest countries.

A particularly promising strategy is to create social institutions of peers directed at nurturing careers that will contribute to development. For example, the African Academy of Sciences could initiate a prestigious Fellows Program for study outside Africa, not only in the United States and other developed countries but in other countries of the South as well. The purpose of the program would be to create a mobile intellectual reserve, fluent in science and technology at the world level and dedicated to development within Africa. Current and former holders of the Fellowships would meet periodically to exchange experiences and undertake cooperative projects and studies. A similar model might apply in Latin America and elsewhere. Comparable programs operated by the Rockefeller Foundation and other organizations were successful in earlier decades in building leadership cadres. Funding of such programs might be shared by private foundations, bilateral and multilateral aid agencies, and the governments of the countries of origin. Both sending and receiving nations must have a stake in the program. The program of Hubert Humphrey Fellows provides one base on which to build.

Organizations in the United States such as the Fogarty Center of the National Institutes of Health and the National Science Foundation, as well as private groups such as the American Association for the Advancement of Science, could do far more to create a sense of community and shared purpose among students from developing countries studying science and technology in the United States. Working groups of students from different institutions might be formed to address various topics, with the objective of creating a larger and more effective network of scientists and engineers dedicated to science and technology for development, whether the individuals return to their home countries, stay in the United States, or pursue careers elsewhere. At present there is *no* private or governmental organization in the United States providing a focal point for issues concerning foreign students from developing countries.

The recent changes in international relations offer even greater opportunities for the United States to apply its unique assets in higher education to development. The end of the Cold War has also broken barriers to cross-national scientific exchange, student travel, and educational cooperation. It lessens the political and polemical pressures that often weigh upon students. U.S. universities individually and collectively and with potential partners in government and the private sector should use this period to consider new and enhanced roles in development.

Source of data: Institute for International Education and UNESCO.

Table 5. Leading Countries of Origin of Foreign Students in the United States, 1990–1991

Rank	Locality	Students
1	China	39,600
2	Japan	36,610
3	Taiwan	33,530
4	India	28,860
5	Korea, Republic of	23,360
6	Canada	18,350
7	Malaysia	13,610
8	Hong Kong	12,630
9	Indonesia	9,520
10	Pakistan	7,730
11	United Kingdom	7,300
12	Thailand	7,090
13	Germany	7,000
14	Mexico	6,740
15	Iran	6,260
16	France	5,630
17	Singapore	4,500
18	Greece	4,360
19	Jordan	4,320
20	Spain	4,300
21	Philippines	4,270
22	Turkey	4,080
23	Brazil	3,900
24	Lebanon	3,900
25	Nigeria	3,710
26	Saudi Arabia	3,590
27	Colombia	3,180
28	Israel	2,980
29	Venezuela	2,890
30	Peru	2,800

Source: Institute for International Education, *Open Doors 1991–92.*

to reduce the obstacles that impede national leadership by the United States in cooperation for development. Current U.S. Government programs need to be freed from the many fetters that bind them if they are to keep pace with the new and complex problems and opportunities facing development. And the public sector needs to fit its own programs better into the work of active partners from the private sector and nonprofit institutions. Box 7 provides an illustration of an area of development program cooperation, tuberculosis control, which represents just such a partnership of interests, expertise, and management.

DETERMINANTS OF CURRENT GOVERNMENT PROGRAM CONTENT

For the past several decades, the substance of the development assistance programs of the U.S. Government has not been governed by such a compact set of criteria. Rather, it has been largely set by three considerations: earmarking of appropriations, the idea of "basic needs," and physical geography.

EARMARKING

Congressional earmarks, or "functional accounts," reserve monies for problems and initiatives favored by particular domestic (i.e., U.S.) constituencies and interest groups. This earmarking process has created *de facto* priorities and has driven the program content of cooperation for development by requiring specific amounts of resources to be allocated to specific sectors—for example, agriculture, child survival, and women's programs. Similarly, in the security assistance program, specific amounts of money have been allocated by Congress to specific countries.

About 85 percent of the current U.S. foreign assistance budget is locked by these processes into specific sectoral programs or countries. Only the Development Fund for Africa and funds for Eastern and Central Europe are not explicitly preallocated by Congress to a specific country or to a specific sector. Cooperation for development is driven by a complex accretion of legislative preferences: neither Congress nor the Executive Branch has any universal vision of an overall program or development strategy. The U.S. Government has fallen into a classic social trap. In the desire to solve particular problems, the nation has created a whole that is much less than the sum of the parts. In fact, the current set of parts can never constitute a whole.

DATED DEFINITION OF NEEDS

Since legislative action in 1973, the major priority for U.S. foreign assistance has been for "basic needs," defined as food and nutrition, population control and health, and basic education. This formulation has been applied globally, to developing countries of every kind and condition. The defini-

Box 7. Control of Tuberculosis

About 1.7 billion people, or 33 percent of the world's population, carry the TB pathogen. Every year, 8 million people develop clinical disease. Untreated tuberculosis has a fatality rate of over 50 percent, with the heaviest toll among young adults—the parents, leaders, and workers of society. It is estimated that one quarter of avoidable adult deaths (ages 15–59) in the developing world are due to tuberculosis. More people die from tuberculosis each year than from malaria or measles.

Estimated Annual Risk of TB Infection, New Cases, and Deaths from TB for the Developing World, 1985–1990

Area	Annual risk of TB infection (%)	New cases per year	Deaths per year
Sub-Saharan Africa	1.5–2.5	1,313,000	586,000
North Africa and Western Asia	0.5–1.5	323,000	91,000
Asia	1.0–2.0	5,102,000	1,825,000
South America	0.5–1.5	356,000	111,000
Central America and the Caribbean	0.5–1.5	185,000	80,000
Total developing world		7,280,000	2,692,000

Source: B. R. Bloom and C. J. L. Murray, "Tuberculosis: A Commentary on a Reemergent Killer," *Science*, 257:1055–1063, 1992.

Two new problems, HIV and drug resistance, worsen the situation. Tuberculosis exemplifies both the problems and the potential for problem solving in a highly interconnected world.

Research and Control. There is need for research programs as well as for control programs. Both efforts must focus on developing national capacity for self-sustaining national tuberculosis control programs. Capacity building for control and research activities will require training, learning by doing, exchanges and visits from the scientific community, and a range of services, including information support and conferences, that will put health workers in individual countries in touch with the broader community dedicated to combating this disease.

Governments will need to develop the public health policy framework within which control can succeed. Private voluntary organizations at the community level will be critical to the outreach that brings control programs to those in need. Private companies, academic centers, and government laboratories will conduct the research and development on cure and control, and products will be manufactured by the companies.

Box 7. *(continued)*

Programs should be affordable. It is estimated that 80 percent of tuberculosis cases in developing countries could be cured for approximately $150 per cure.

Global Problem, Global Programs. Tuberculosis illustrates the value of a global perspective, with pluralism of response, and the essential role of science and technology in overcoming obstacles to implementation. However, it also illustrates some of the anomalies of applying science to development.

The priorities for disease control and health research worldwide are determined by the preferences of industrialized countries. Control of tuberculosis should rank among the most important health priorities in developing countries. Until recently it has not, mainly because the disease had ceased to be important in industrialized countries. The substantial research capability in tuberculosis was phased out in industrialized countries, eliminating a critical resource for training and research that could meet the continuing needs for tuberculosis research in developing countries.

Renewed recognition of the importance of tuberculosis was not based on epidemiological evidence gathered by developing countries in order to establish their health priorities. Instead, an international commission to assess the priorities for health research in support of development focused attention on the neglect of tuberculosis; the commission was spearheaded by independent-sector foundations in cooperation with multilateral agencies.

At the same time, interest in tuberculosis in industrialized countries has been growing because of its reemergence as a health problem at home. Immigrants have arrived with active disease and it is associated with AIDS; these two factors, along with the emergence of new disease-resistant strains and the difficulties of ensuring compliance with the treatment regime, mean that cities like New York face daunting problems with TB. In short, the problem of tuberculosis illustrates the interdependence of all countries in the fight to eliminate disease.

The scientific agenda for overcoming obstacles to control is also global. For example, much work is needed to improve tools for case detection of reactivation of tuberculosis, to examine the causes of drug resistance, and to simplify the treatment regime in order to enhance compliance. Not all advances in the control of tuberculosis originate in the sophisticated institutes of industrialized countries. Indeed, the field studies carried out in Bangladesh by a nongovernmental organization, the Bangladesh Rural Advancement Committee, have resulted in an incentive system to patients and health workers that has nearly doubled the completion rate for tuberculosis treatment to a level close to 90 percent. This research is an example of many contributions from developing country organizations to the advancement of knowledge for worldwide benefits.

Tuberculosis reflects the value of a global perspective and the benefits of a pluralistic response from governments, nongovernmental organizations, multilateral agencies, the independent sector, the science and technology community, and industry, with its special responsibility to develop better and more affordable tools for intervention.

tion of "basic needs" has not altered in twenty years, despite striking changes in the world and in the status of many countries.

Much of the original definition of "basic needs" continues to be sadly relevant for the world's poorest countries. Even in these countries, however, basic needs programs have had little leverage on economies overall, unless they were accompanied by rigorous mechanisms for replication (and this has been rare).

Moreover, economic growth, employment, capital markets, technical skills, information technology, telecommunications, energy, environmental quality, democracy, and freedom—all of these requirements for continued economic and social progress in any country are also "basic needs." Such crucial ideas fall outside the legislative definition of "basic needs" and thus often cannot be part of U.S. cooperation for development, notwithstanding the legitimacy of the need or the desire of the development partner.

Obsolete Geography

Reflecting former concerns about the regional spread of Communism and other political threats and traditions in international relations, most of the U.S. Government's development programs group nations simply by geographic location, rather than by criteria concerning economic or social condition. The substance of the work specified in legislation is thus arrayed against geography when it comes to the design of programs. Although this arrangement allows program managers to share and learn from some common cultural elements, it discourages sharing and learning from opportunities and approaches that are common economically. It is, for example, difficult to cross-fertilize programs in Hungary, Thailand, Mexico, and Morocco, because they are classed as "different" geographically rather than "similar" economically.

Resulting Mismatch

These three features—highly detailed earmarking of appropriations, dated definition of "needs," and obsolete clusters of nations based upon geography—combine to undercut the effectiveness of U.S. cooperation for development in several ways.

The specification of expenditures for particular countries and sectors drives programs to respond to legislative requirements rather than to country condition or level of need. A mandatory level of expenditure for child survival, for example, leads programs to look for projects for child survival in

Rotary International parade promoting immunization, Madras, India.
(Photograph by Christopher Warren.)

all countries, even if malaria, tuberculosis, inadequate water supply, or even cardiovascular disease are the overwhelming killers in many populations.

Further, because so many detailed earmarks are made, only a small amount of money is available for the few flexible "functional accounts." With the exception of the traditional fields of agriculture and population, the tendency is to specify ever more narrowly how money should be spent even as the potential scope for action expands. For instance, the account for "private sector, energy and the environment" is the third smallest in the budget, at about $150 million (compared with agriculture, at about $500 million). Yet this account is the main source for funding projects intended to address an enormous range of emerging development problems. No matter how compelling an energy or environmental problem in a country, it cannot be addressed with the funds allocated to agriculture or population.

Dynamic adaptation to major opportunities is thus constrained by either budgetary compromises or inertia. A 5 percent increase in functional accounts or country budgets does not allow programs to respond to the sudden, or even less than sudden, appearance of new problems or to the recognition of new potentials. The detailed nature and level of U.S. cooperation for development is largely predictable, while, unfortunately, the problems to be addressed often are not; this has been illustrated powerfully by the rapid changes in health and environmental needs, and in the new challenges for privatizing the inefficient state enterprises, as seen in Eastern Europe and the former Soviet Union. As a result, the U.S. Government has money to spend on "last year's problems," problems that, in many countries, are no longer (or will soon no longer be) the critical paths to economic and social progress. Needless to say, the money is always spent, even if the problem no longer is as important as others.

NEW APPROACH

Increasingly, an entirely new approach is needed. From the perspective of U.S. interests, the critical international boundaries are economic and social, not geographical. The dominant concerns for America's own future are stable growth in the world economy that does not compromise the global environment for future generations, free patterns of trade and open investment, increased participation of all nations in trade and economic growth, and slowing of world population growth. The distribution of goods and services around the world and how they are produced is now as important to the security of the United States as the array of arms was four decades ago. Given the rising importance of global economic performance, the United States must update the substance of its development programs to recognize the diversity of conditions among developing countries and the evolving problems countries face as their economies and societies develop. Partnerships in global development require effective U.S. national and governmental capacity to cooperate with the full spectrum of countries and on problems that cut across national borders.

FULL SPECTRUM OF PARTNER COUNTRIES

Thirty or forty years ago, the condition of recipient countries was, though not uniform, similar in key respects. Economies dominated by agriculture and natural resource exploitation, low literacy, short life expectancy, deficient infrastructure for water, and high birth rates characterized the majority of developing nations. Except for the remains of the colonial heritage, most were tied only marginally to the international trading system, and they lacked significant domestic or foreign private investment. The administrative apparatus of the modern nation state was absent in many cases.

Today, there is much more diversity in the development of nations. Some countries move consistently forward; some make progress slowly and episodically; some remain overwhelmed by poverty and conflict. The diversity is less defined by geography or region than ever before. Thailand, India, Brazil, and Morocco resemble each other more than they do their geographic neighbors Cambodia, Bangladesh, Bolivia, and Yemen.

To cooperate for global development, partnerships must form across an economic spectrum including advanced developing countries, middle-tier countries, and the poorest countries. Moreover, a single country may itself display many different levels of development, both regionally and by economic sector.

- *Advanced developing countries* are those countries whose trajectories of economic growth and social progress are positive and stable and whose greater integration into the global economy is a realistic near-term prospect. These countries are also often characterized by rapid learning in the society of new skills and new ways to solve problems. Among such countries are Thailand, Hungary, Mexico, Brazil, and Costa Rica.
- *Middle-tier countries* are those countries that have experienced significant progress in economic growth and social evolution, but whose pace is slower and whose path includes more switchbacks and barriers. In these countries, the ability to overcome the barriers is weak. Such countries include Egypt, Poland, Pakistan, Jamaica, and Indonesia.
- *The poorest countries* remain challenged by the most fundamental problems of poverty and instability. For these, the engines of economic and social progress remain unfueled, and the track onward is steep and long. In these countries, the capacity for self-reliance in development is extremely limited. There has been little diffusion of science and modern technology. Many countries in this category are in Africa, and the list also includes such nations as Laos, Cambodia, Bolivia, Haiti, Afghanistan, and Albania.

Adaptive Programs

Because all developing nations were perceived as sharing similar conditions four decades ago, U.S. programs for aid emphasized a few core concerns that were broadly applicable to many regions. Agricultural development took first place in U.S. programs, followed by public health, population, and basic education. The substance and approach of U.S. programs were similar everywhere. Today, however, with the varied improvements around the globe and the emergence of a new group of partner countries in Eastern Europe and the former Soviet Union, the mandate is for flexibility rather than for a few centrally chosen formulas.

Education provides an example. U.S. development programs support education only when it is targeted at basic literacy. The congressional earmark specifies that funds for education be expended on basic education. In countries without this specific problem, education programs are generally not supported. But the diversity of conditions in developing nations, when placed against this stricture, brings undesirable results. As literacy rises, support for education, the essence of capacity-building, declines. Opportunities to respond to educational need at higher and equally important levels are lost. Rather than build on successes in education, America walks away from them. The building of capacity that can sustain development, enabled by successes in basic education, cracks and crumbles.

Adaptive programs, driven by country condition rather than central mandates, are required to correct such missteps. Again, education provides an example. Where literacy remains the critical hurdle, U.S. cooperation should retain its present form. Where secondary education or technical training is the emerging problem, however, cooperation for development should be capable of an appropriate and equally vigorous response. And where strengthening university training in science and engineering fits the evolving condition of a partner country, this too should be an option. Cooperation for development should be based on the premise that there are no limits to useful learning. Indeed, if the developing societies are to be competitive in tomorrow's international economy, just such flexibility in all donors' educational programming will be needed. It is estimated that, of new jobs created in the 1990s in industrialized countries, 49 percent will require a minimum of 17 years of formal education, while at most a third will be appropriate for workers with less than 12 years of schooling. As a recent report notes, "the implications . . . for developing countries are stark."[53]

No More Top-Down

A blunt instrument of top-down management, the traditional earmarking is absolutely contradictory to the concepts of development cooperation based on partnership. The U.S. Government should adopt dynamic definitions of needs that correspond to the changing conditions in partner countries and in groups of countries (grouped by mutual interests). The substance of cooperative relationships must be able to range over such areas as strengthening private investment in industry; increasing trading capacity; reforming the legal and regulatory environment as it affects investments (e.g., intellectual property and commercial law) for local entrepreneurs and international investors; education and management training at all levels; strengthening of research and development capacities, especially in sectors likely to generate future employment; alleviating suffering and making essential improvements in the human condition; and transferring the skills and key resources critical to stabilizing the welfare of the most vulnerable populations.

CONCLUSION

Rapid and widespread change logically requires that the United States unbind its cooperation for development and adapt it to the new landscapes of political, economic, and technological opportunities. Cooperative development programs must more effectively balance growth with equity, management with participation, large-scale with small-scale endeavors, global campaigns with local needs, and the establishment of rules and norms with investment in bricks and mortar. Partnerships must recognize the complementary nature of private markets and government policy, the importance of pluralism and experimentation in efforts to promote economic and social welfare, the role of the individual in economic progress, and the interdependence of growth and environmental integrity.

The United States must pursue initiatives to alleviate human suffering, employing what we already know while retargeting longer-term programs to tap the vast potential of science and technology in order to uncover new means for solving the most difficult problems.

The evolving content of cooperation simply cannot be expressed within the old forms of "aid." In particular, the current authorizations for, and the obsolete organization of, U.S. development assistance impede effective action. How to embrace diversity and encourage flexibility and what must change to achieve new objectives are the subjects of the next chapter.

4
HOW: ORGANIZATION, DECISION MAKING, AND RESOURCES

The institutions and decision-making processes set in place for development more than a generation ago are ill-suited to evolving U.S. interests and the diversity of the nations with which the United States must now cooperate. This part of the report addresses the fundamental changes needed in the arrangements for U.S. cooperation for development. How to mobilize the full national capability, including the private sector and nongovernmental organizations, is addressed first. Next, the report discusses changes in the federal government, including the White House, Congress, and Executive Branch agencies, particularly the U.S. Agency for International Development.

Then the report comments on facets of the multilateral system for cooperation in development in which the U.S. participates, including the World Bank and the United Nations system. For the future, a rising wave of multilateral action will be significant, even as the key roles for bilateral partnerships remain essential. Finally, the report addresses the question of resources.

Beginning with the Gardner Report of 1964, numerous studies have sought to improve the substance, organization, and conduct of U.S. foreign assistance. These studies contain many useful insights that the Task Force has taken into account.[54] Several of the changes that the Task Force urges are consistent with earlier recommendations that were not acted upon because the time was not yet fully ripe; now, the new geopolitical environment offers much greater prospects of success.

HARNESSING THE FULL POWER OF PLURALISM

The size of the total U.S. national effort for partnerships in development is enormous. And yet none of the participating sectors, whether the public sector, the private sector, or the independent sector, appears to be fulfilling its potential. Moreover, only in rare cases are the sectors effectively allied with each other. The government, the private for-profit sector, the private voluntary organizations, universities, and the foundations of the independent sector all need to improve their ability to network internally and to work together across institutional boundaries. Greater effectiveness would undoubtedly result from such coordination.

Independent Sector

In the independent sector the private voluntary organizations (PVOs) are among the most important participants in development cooperation. American PVOs are active throughout the former Third World, and are increasingly involved in the newly democratizing countries of Eastern and Central Europe and the Commonwealth of Independent States. Both as participants in programs funded by the U.S. Government and as independent actors, they play a significant role in social services, training, housing, health, and agriculture. Moreover, American PVOs have sparked the formation of thousands of counterpart local private voluntary organizations within partner countries. This network is becoming an important source of private initiative and self-reliance throughout the developing world.

While diversity and independence are precisely the strengths of the independent sector, the effectiveness of the sector could be greatly enhanced by stronger incentives for networking, joint meetings, and consultation. In fields such as hunger and biodiversity, the benefits of such cooperation have already been amply demonstrated in the past few years. Shared databases and support services demonstrate the value of practical coordination. The

Interaction Council and the Citizens Network for Foreign Affairs are nongovernmental organizations (NGOs) whose cross-cutting efforts exemplify how much strength remains to be tapped in the independent sector.

■ **The Task Force recommends that leading organizations in the independent sector concerned with cooperation for development using S&T explore mechanisms for regular exchange of information and extension of voluntary networks to address common concerns.** The mechanisms developed should be sharply problem-oriented so that participants can sense their shared mission and fulfill action plans. Few NGOs concerned with development have focused on the wide-ranging science and technology they use and need, or on their individual or collective capacity for substantial research and development. Jointly sponsored analysis, and even applied research, would buttress the longer-range programs.

Private Sector

The U.S. commercial private sector is also a powerful contributor to global development. The annual flow of direct U.S. foreign investment to developing countries is about $9 billion, greater than the annual flow of U.S. Government bilateral economic assistance.[55] Such private investments appear to be increasing as opportunities improve in many countries. Moreover, private trade with developing economies is also critical to the economic health of the United States. Developing economies purchase 35 percent of all U.S. exports, accounting for about 2.5 million jobs in the United States.[56]

American companies are also regular philanthropic contributors to U.S. development programs. More than 80 U.S. companies have contributed pharmaceuticals and medical supplies to the Commonwealth of Independent States through AID programs, implemented by Project HOPE.[57] American corporations regularly support U.S. private voluntary organizations with both cash grants and in-kind services and materials for their programs in developing countries.

Hence, both as an engine for economic growth and expanded trade, and as a generous partner in philanthropy, the American commercial sector occupies a pivotal role in the direction of future global development.

The U.S. private sector has several mechanisms that promote valuable exchanges of information among its leaders and enable coherent action on problems of common concern. These include the Conference Board, Committee for Economic Development, Economic Club, Chamber of Commerce, and Industrial Research Institute; such organizations often include key people from organized labor and from various special-issue organizations. Historically, such organizations have taken little interest in growth and change in the economically less-advanced countries of the world. But future opportunities for global trade mean that this attitude must change.

■ **The Task Force recommends that the major organizations that link high-level U.S. business and labor executives for exchange of ideas on economic policy form standing study groups and action-oriented panels concerned with long-range global development and the role of U.S. private enterprise.** Environment and sustainable development provide an example of an area where U.S. firms might agree on operating principles and actions to be taken in developing regions, as outlined by the Business Council for Sustainable Development in its 1992 report *Changing Course.*[58] The World Bank and other multilateral and international financial institutions should also encourage conditions congenial to private initiatives.

National Action Roundtable

Ultimately, government, private, and independent sector efforts will be effective only as parts of a meaningful and balanced triad. For example, cooperation for development by the U.S. Government must deepen support for private voluntary organizations — in the U.S. and especially abroad — because they are the grass-roots sources of creativity and continuity in pushing for humane innovation and responsible government. Equally, U.S. Government programs must forge closer links with corporate America. Private commerce and entrepreneurship can play an important role not only in the charitable aspects of development programs, but also in creative program planning, leading to self-sustaining commercial development. With their global networks and sophisticated scientific and managerial capabilities, U.S. corporations can also analyze developing-country economic and policy conditions, and advise about the timeliness and the prospects for success of various strategies. Finally, to carry out truly global campaigns on urgent problems such as tuberculosis or deforestation, the cooperation of *all* sectors is required.

■ **To foster creative cooperation among these diverse sectors, the Task Force recommends the creation of a National Action Roundtable for International Development.** The Action Roundtable would include representation of concerned, expert leaders from the executive and legislative branches of the federal government and from the private and independent sectors. The purpose of the Roundtable would be to review the evidence on trends and then catalyze the creation of specific intersectoral coalitions to address particular problems. Each proposed solution would be clearly in the international interest, and each must be justified in a convincing way to the American public.

U.S. GOVERNMENT

Within the U.S. Government, the operative rationale and most actual programs for development are three or four decades old. For the past twenty years, there has been only episodic concern at the highest levels of government regarding American strategy for international development. The purposes defined and the mechanisms set in place after World War II have been considered sufficiently appropriate and productive to survive attempts at reform. The Task Force believes the *status quo* in the government is no longer sufficient. Most important in achieving fundamental changes will be leadership. Given the entrenched interests, institutional inertia, and accretion of organizational complexity created over four decades, that leadership must come from the White House. At the same time, Congress must act to reform the legislation that governs cooperation for development and the mechanisms for oversight of "foreign assistance" programs.

THE WHITE HOUSE

■ **Recognizing the need for leadership, the Task Force recommends that the President articulate principles and long-range priorities for cooperation with the entire range of developing countries.** Presidential guidance must be based on the best evidence about U.S. interests in, and any new circumstances within, developing countries. The strategy must include not only the poorest countries, but those in the middle tier and more advanced levels. The central roles of science and technology in almost every program should be rethought, and sophisticated advice sought from many sources. Particular attention should be given to emerging democracies and countries recovering from internal conflicts. Guidance should provide over-arching policy with a few clear goals toward which cooperation for development would be targeted, a timetable for reaching the goals, and a means for periodic review of progress at the cabinet and presidential levels. In order to provide such guidance, the President should order an intensive review by all relevant federal agencies of their activities in development.

The Congress

■ **The Task Force recommends that, concurrent with the Presidential review, the Congress initiate consultations, studies, and hearings that will lead to major reform of "foreign assistance" legislation and oversight.** Current legislation prevents all but marginal improvements in cooperation for development. The Foreign Assistance Act of 1961 has now been amended over 70 times and contains 33 objectives.[59] AID itself has identified 75 equal priorities for American development programs.[60] American government programs must respond to over 100 "most important" goals.

Adhering to and reporting on such complex congressional direction is onerous. Present legislation specifies 288 reporting requirements; this leads to over 700 congressional notifications each year.[61] The four most significant congressional committees for development, Foreign Affairs and Appropriations in the House of Representatives and Foreign Relations and Appropriations in the Senate, are joined by 17 other committees and 20 subcommittees with explicit authority to become involved in U.S.-supported activities in developing countries.[62] Having more than forty overseers is leading to the epitaph: "died of extreme accountability." This pattern must change.

Excessive Earmarking and Burdensome Bureaucracy

Over 90 percent of the budget of Economic Support Funds and more than 60 percent of the Development Assistance budget are explicitly allocated by Congress either to particular countries or to development sectors.[63] This degree of earmarking hampers professionals and managers responsible for U.S. bilateral cooperation for development in their efforts to make programs respond to changing global conditions. The excessive earmarking of both program substance and levels of funding allocated to specific countries creates rigid and ineffective programs and budgets. (See Box 8, which shows the regional distribution of AID financing, and illustrates the impact of military and geopolitical priorities.) Excessive earmarking is often counterproductive, placing the United States in the role of "central planner" rather than "partner" of developing countries.

The combination of complexity of authorization with burdensome reporting requirements also creates the need for a legal and managerial superstructure in U.S. Government cooperation for development. This superstructure then consumes many of the resources that might otherwise be spent in pursuing development itself. The government superstructure in turn generates the need for a matching superstructure in contracting organizations at home and abroad.

Box 8. Evolution of U.S. Assistance by Region, 1946–1989

The chart shows the real value of total U.S. foreign assistance (economic and military) over 40 years expressed in 1989 dollars by region. There were three major shifts in regional emphasis during this time:

- 1946–1952: Europe was the dominant recipient. Total assistance averaged $32 billion per year.
- 1953–1974: Asia was the prime focus. Total assistance averaged $22 billion per year.
- 1974–1989: Israel and Egypt have been the primary recipients. Total assistance averaged $16 billion per year.

There were spurts in aid to Latin America associated with the Alliance for Progress (1962–1967) and in the 1980s to Central America. There was a 40 percent reduction in aid to Latin America between fiscal year 1985 and 1988 resulting from increased U.S. budget deficit pressure. Assistance to Africa, which began to grow in 1976, also suffered a major cutback of 55 percent between fiscal 1985 and 1989.

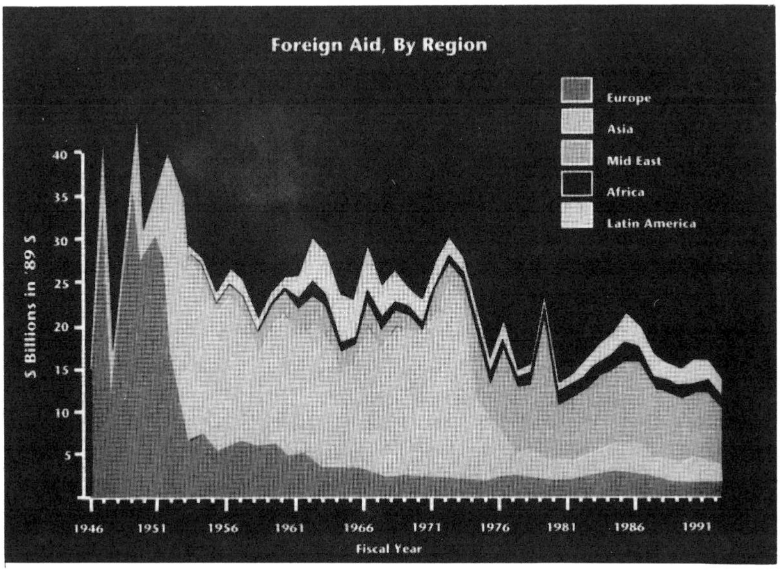

Source: President's Commission on the Management of AID Programs, *Report to Congress—An Action Plan*, George M. Ferris, Jr., Chair, Washington, DC, April 1992.

Less Micromanagement, More Development

Lessening micromanagement can result in creative programs of American Government cooperation for development. Recently Congress provided assistance funds to the newly democratic states of Eastern and Central Europe outside the earmarking process. In turn, this meant that, with neither sectoral nor country entitlements, U.S. "foreign assistance" could be targeted *both* at the most appropriate problems in each country, *and* at those countries that had made the greatest progress in instituting economic and political policy reforms. This made for more effective partnerships. (Box 9 briefly addresses the complex issues of "conditionality.")

Exemption from earmarking in Eastern Europe has led to American support for programs that fit local circumstances and are jointly identified by the United States and the partner country as priorities. These programs range from traditional public health and education projects in poorer countries such as Romania and Albania to transformation of banking, securities markets, medical systems, energy production, and environmental protection in more advanced countries such as Hungary and the Czech and Slovak Republics. Exemption from earmarks thus allowed Eastern European programs to be defined and driven largely by demands in the field. It has also allowed more rapid identification and execution of development initiatives and reduced the time for project design and approval by more than half. Moreover, the disbursement rate for approved project funds in European programs is twice that of the rest of AID.

Another example of the benefits of substantive and organizational flexibility in U.S. programs is the Development Fund for Africa. The Fund, although earmarked as a line-item in congressional legislation, is not further restricted. Its $800 million in project resources are allocated throughout Africa on the basis of performance. Countries that turn toward democracy and liberalize their economies can obtain added resources from the fund, beyond the functional accounts and country budget levels earmarked by Congress elsewhere. The additional fund monies are not specified by sector; they are applied to the development problems that the recipient governments and U.S. representatives agree are most critical.

Not every program in Eastern Europe or supported through the Africa Fund is perfect or perfectly timed. But the flexibility results in better partnerships, adapted to national conditions and quicker to respond to needs and opportunities.

Three Critical Elements

There are three critical elements of legislative reform: stringently limited earmarks on appropriations, allowing flexibility in programs; a reasonably

Box 9. The Problem of Conditionality

The past forty years of U.S. foreign assistance has seen much debate about the appropriateness of making such assistance "conditional" upon various circumstances or policies of the recipient country. Indeed, current legislation raises a number of barriers to the provision of assistance, most premised not on national economic policies but on the political conditions and structures within a country. Sweeping conditionality, the requirement that every country meet some universal economic or political policy standard, is a tempting parameter for foreign assistance.* Certainly, a few limitations on partnership must be set—that a partner honor human rights, international law, individual freedom, and the like. During the next few years, some observers argue that conditions should be set for the reduction of military expenditures or for agreements about arms control. Set too narrowly, however, conditionality threatens the flexibility needed by America if it is to assist a still unsettled world, as well as undermining the very concept of "partnership," which calls for the United States and its development partners to address problems of economic policy and social development together, collegially, and over the long term.

* See C. J. Jepman, *The Tying of Aid*, OECD, Paris, March 1991.

limited set of objectives toward which cooperation for development is to be applied; and measures of effectiveness against which cooperation can be held accountable. The Task Force notes the recent Executive Branch proposals to rewrite the statutory base for foreign assistance. These proposals are broadly consistent with the findings of a bipartisan congressional task force that reported in February 1989[64]:

> Foreign assistance is vital to promoting U.S. foreign policy and domestic interests, but the program is hamstrung by too many conflicting objectives, legislative conditions, earmarks, and bureaucratic red tape. . . . The present system is unworkable and increasingly irrelevant. . . . U.S. foreign assistance needs a new premise, a new framework, and a new purpose to meet the challenges of today.

The Task Force finds great merit in these efforts and urges early action to institute substantial reform.

Convergence

The presidential review and the legislative reform should be pursued collaboratively and with full exchange of views. The efforts of the executive and legislative branches should converge during 1993, enabling the acutely needed new legislative framework to be enacted.

Coordination of Executive Agencies

Numerous U.S. federal departments and agencies are involved in development.[65] The Agency for International Development, addressed separately below, is the government's primary foreign assistance organization. Among the other significant agencies are the Department of Agriculture (DOA), the Department of Health and Human Services (DHHS), the Overseas Private Investment Corporation (OPIC), and the commercial agencies. This last category includes the U.S. Special Trade Representative in the Executive Office of the President, the Export–Import Bank, and the International Trade Administration (ITA) of the Department of Commerce. The Department of the Treasury is responsible for U.S. participation in the World Bank, the regional development banks, and other international financial institutions. The Department of Justice houses the Immigration and Naturalization Service and the Drug Enforcement Administration.

The Peace Corps

Within the federal government special mention must be made of the Peace Corps. Since its founding in 1961, more than 120,000 Peace Corps volunteers have completed assignments in developing countries and have returned to the United States with skills and perspectives that have not only been important in their own lives but that have often led them to continue their commitment to global development through their choice of profession. This network of existing and former Peace Corps volunteers represents one of the most important sources of support for and expertise in development cooperation.

Broad Federal Involvement—Benefits and Difficulties

Of the non-AID federal agencies, and with the obvious exception of the Peace Corps, the Department of Agriculture has the most significant overseas presence. It has 86 Foreign Agricultural Service Officers in U.S. embassies in the developing world, provides extensive technical assistance to AID itself, and is a major participant in the PL480 program, which transfers excess U.S. grain stocks to developing countries in exchange for local currency. Another agency with a strong presence abroad is the Department of Health and Human Services (DHHS). Critical elements of DHHS are the Centers for Disease Control and some of the National Institutes of Health. These are renowned scientific centers whose expertise, technology, and advice are sought throughout the developing world.

The critical benefit of broad federal government participation in cooperation for development is access to the networks of science and technology that each specialized agency brings, whether in agriculture, energy, health, telecommunications, transportation, space, natural resource management, or environment. The value added in broad government participation is not in substituting the bureaucracies of other domestic agencies for that of AID. The value is the expertise and contacts of the individuals whose involvement can ensure quality and reliability of programs. Moreover, broad participation makes possible a continuous coverage of needs as nations evolve economically through different levels of development. At earlier levels, a developing nation may be more likely to have close relations with AID, while at more advanced levels the center of gravity in partnerships may shift to the commercial agencies and to the National Science Foundation, for example.

Despite the participation of many federal departments and agencies in foreign projects, there is often little or no coordination among them. Most requests for interagency cooperation in technical assistance are subject to detailed and project-by-project negotiations between AID and other agencies after AID has received its appropriation. In fact, federal agencies often see their roles in development as competitive with AID, a battle to gain added resources from the development appropriation. The transfer of development resources to domestic agencies allows them to charge part of their overhead to the development budget rather than to domestic operations. Under these conditions, the commitment of federal agencies other than AID to development rises and falls erratically in response to internal pressures and the personal vision and interests of senior department managers, and there is little coherence to the overall effort.

An interagency coordinating institution created in 1979, the International Development Coordination Agency (IDCA), exists on paper but was only briefly operational. Recently, interagency coordination has been attempted more successfully in the highly visible Eastern Europe program by a senior Coordinators' Group chaired by the State Department and composed of high-level Treasury, AID, and White House representatives. This group regularly reviews the program and ensures communication among participating agencies.

The Presidential Review

One outcome of the presidential review proposed earlier should be a precise definition of an effective coordinating mechanism for federal programs in development. To conduct the presidential review, and to carry on subsequent coordination, participating federal agencies must designate focal points for development issues, typically an assistant secretary for international affairs, or for policy, or for science and technology. Continuing follow-up at such a level after the review is completed will then make coordination more authoritative. Such designation will also help ensure that concern is institutionalized within agencies rather than remaining a function of the personal interests of temporary top appointees. It may even help ensure that the top management team in all key agencies always includes at least one individual who is recruited in part for his or her expertise and commitment to development.

A nagging and difficult question is the relation between military and other forms of assistance. Whatever the merits of past proposals and practices for separating military and development cooperation, future international security will rest more on building programs for peaceful change and development than on deterrence. If, as the Task Force hopes, during the 1990s there is a marked shift away from military assistance as a lever in international politics, assessing development opportunities and relating them to American foreign policy and security goals will become more important.

Dealing with the Problems

▪ **The Task Force urges the strengthening of the means for interagency cooperation in international development.** Accessing expertise across the federal government for U.S. development programs, yet ensuring that coordination rather than bureaucratic competition characterizes that process, probably requires the leadership of a disinterested *nontechnical* department. Because renewed cooperation for development is above all tied to the goals of American international relations, the Task Force recommends that the State Department take the lead role in integrating the government's programs and strategies for cooperation for development as well as meshing it with foreign policy goals. Unfortunately, development has been marginalized as a goal of U.S. foreign policy, and it needs to be brought back into the mainstream. State Department leadership in interagency coordination is one way to help achieve this.

U.S. Agency for International Development

The Agency for International Development, an independent agency reporting to the Secretary of State, is the U.S. agency with the longest history of involvement in foreign assistance.

While much about U.S. development cooperation is in need of reform and reinvigoration, the United States has ample grounds for pride in many major accomplishments of the past. These accomplishments include the eradication of smallpox, the success of the Green Revolution, the pioneering of population planning, the introduction and widespread adoption of oral rehydration therapy, and the spreading of economic reforms. Such achievements, made possible with the support of dedicated American staff and substantial resources, should not be forgotten.

Currently, AID has the most significant explicit financial and policy responsibility for U.S. foreign assistance. First, consider the basic facts of funding. AID administers several types of resources. The Economic Support Fund (ESF), totaling $3.24 billion for FY 1992,[66] represents cash transfers, commodities, and development projects for countries of strategic importance to the United States. Egypt and Israel account for about 80 percent of these funds. Development Assistance (DA) resources are more broadly used throughout the developing world. The DA appropriation for FY 1992 was $2.2 billion.[67] More than 60 percent is earmarked by Congress for specific sectors, above all for agriculture, population, child survival, and education. Operating expenses, costs for auditing and the Inspector General, and pension management add another half billion dollars to AID's budget.

As for staff, in January 1991, AID's employees worldwide totaled slightly more than 11,000, of whom only 30 percent were American foreign service or civil service personnel with full-time, permanent tenure.[68] The majority of AID's full-time staff (61 percent) are contract personnel hired to carry out staff functions. This proportion has grown over the past 25 years, as AID's permanent personnel allocations have been reduced. Over half of AID's total staffing, and nearly three-quarters of its field staff, are non-direct-hire foreign nationals serving on a contract basis, most in AID missions in their home countries.

Organizational Challenges

In its position on the front lines of foreign assistance, AID reflects the effects of 30 years of declining public support and increasingly restrictive legislation. AID faces severe challenges as an organization. The combination of popular disillusionment about use of public funds for foreign assistance with the increased congressional control over the details of projects, has reduced the attractiveness of AID for professional careers in science, engineering, and medicine. This is especially true for those needed to address emerging development issues such as energy, the environment, telecommunications, and information systems. Technical experts are vastly outnumbered by lawyers, auditors, inspectors, and administrative personnel.

Moreover, the external efforts to micromanage have caused AID to

Children and math books in Albania. U.S. development assistance helped finance the rewriting and printing of math textbooks, formerly filled with Communist propaganda, for all elementary school students in Albania. (Photograph courtesy of AID.)

centralize and to focus on bureaucracy rather than on field programs. About two-thirds of AID's full-time direct-hire U.S. staff are located in Washington rather than in its field missions close to the sites of actual cooperation. Of the Washington headquarters staff, less than a quarter work in the four geographic bureaus that actually develop and oversee programs and policies in the field.

The ability to implement a project also has suffered. The average time from the identification of a cooperative project to the first flow of funds is over two years. The delay results from an overburdened technical staff, weak staff presence in the field, and the need to ensure that the project responds to a plethora of congressional conditions. The statutory checklist to be filled out for every AID project to ensure conformity with all amendments and congressional requirements is 37 pages long and contains approximately 150 separate requirements for evidence.

Micromanagement has also compromised long-term planning and analysis at AID. With resources and substance tightly controlled by the structure of authorizations, there is little opportunity or incentive for AID professionals to develop information and analyses to identify and track emerging problems and opportunities, or to assess the evolution of problems being addressed by cooperative programs. In turn, the lack of planning and strategy weakens AID's own efforts to present and justify new directions and increased program flexibility to Congress.

Because of the emphasis on expenditure within tightly predetermined program areas and strict reporting requirements, the set of organizations with which AID can contract is highly concentrated. Some years ago, a study showed that the top ten private voluntary organizations received nearly three-quarters of the dollar value of AID contracts.[69] Among universities with S&T relationships with AID, two-thirds of the value of the contracts went to one-seventh of the institutions.[70] The new AID initiative on University Development Linkages represents a constructive, if modest, attempt to strengthen the quality and broaden the range of U.S. and foreign counterpart institutions; joint programs pursue independent ideas about paths to development, especially in capacity-building.

Measures of Success

Finally, what about the results of AID programs? As noted earlier, there have been many positive results over the years. Yet a comprehensive net assessment is difficult. Because there is no generally held view about what the U.S. bilateral assistance program is trying to accomplish, it is uncertain which measures tell the story. If the ability to sustain a program after formal AID participation ends is used as an indicator, there is some disquieting news. In a 1987 study of 212 projects, AID's Office of Evaluation found that only 11 percent had a long-term probability of continuing.[71] A 1989 examination of 62 completed health projects found that "more than half of the projects either had failed before project completion or were unlikely to be sustained following termination of U.S. support."[72] There have been few such rigorous evaluations of clusters of projects. Furthermore, no comprehensive picture of the effects of AID's program emerges from the isolated reviews and anecdotes.

Reform: The Need and the Way Forward

Reforms could improve the organization of AID and in particular its ability to employ science and technology for development. Nineteen recommendations for change to address current problems are offered in the April 1992 Report of the President's Commission on the Management of AID Programs, chaired by George Ferris.[73] The Congress asked for this study, and Box 10 provides a summary of the findings, which reflect both executive and congressional views on the need for change. The Carnegie Task Force's independent appraisals over 1990–1992 are generally consistent with these findings.

Box 10. Key Recommendations of the President's Commission on the Management of AID Programs

Critical Underlying Issues

Redefine the mission and objectives of foreign assistance and enact new legislation
Fully integrate AID into the Department of State
Establish a senior Executive Branch coordinating group for foreign assistance
Limit Economic Support Fund expenditures to political foreign policy objectives

Restructuring Program Management

Establish a Chief Operating Officer for AID
Reduce Washington programs and organizational elements
Standardize planning, information, and management procedures
Integrate budgeting of programs, operating expenses, and personnel
Reduce number of and simplify country programs
Concentrate program resources on private sector economic growth

Improve Personnel Management

Design and implement an effective system for workforce planning
Match recruitment to changing technical needs
More vigorously manage assignment and career development system
Adapt training programs to changing technical needs

Improve Accountability

Develop a performance management system to link employee work plans to evaluation
Make worldwide financial and contract reporting system mandatory
Strengthen internal control review process and link to management planning
Structure programs to reduce private, PVO and university dependence on AID resources
Introduce two-year appropriations cycle for Development Assistance (DA) funds

Source: President's Commission on the Management of AID Programs, *Report to Congress—An Action Plan*, George M. Ferris, Jr., Chair, Washington, DC, April 1992.

Most reform within AID of organization or process can deal only at the margins of the problem. Indeed, the dedicated and competent staff at the agency—whose numbers are shrinking and whose morale is low—need new opportunities to tackle the problems of the 1990s. However, AID remains hostage to decades-old legislation, bound by hundreds of amendments, reporting and contracting requirements, and priorities. Drained of the technical lifeblood of its professional core, AID as an institution cannot be retooled overnight to grasp the future and lead American bilateral development toward the new types and levels of cooperation envisioned in Chapters 2 and 3 of this report. AID is the master neither of its own future nor of the future of the American government's cooperation for international development. That future requires first the changes external to AID mentioned earlier: that the President lead U.S. policy toward a future in which America's security is measured by global prosperity and social progress, even in the face of isolationism at home; and, equally, that Congress attend seriously to negotiating the legislative reforms that are a prerequisite for even small steps toward more effective cooperation for development.

Assuming that the President and the Congress act as suggested, creative and committed leadership at the helm of AID could address present problems and seize the new opportunities. With such leadership, new perspectives and organizational arrangements that emphasize condition and opportunity, not geography, become possible. Invigoration of staff skills, decentralization of authority, insightful long-term planning, far-ranging use of American expertise in science and technology—all can have real impact on AID's functioning. The suggestions contained in the Ferris Report then become powerful.

Without such major reform, there is a strong temptation to bypass existing mechanisms and create alternatives, such as the Sustainable Development Fund (SDF) proposed by the Overseas Development Council.[74] As visualized, the SDF would gradually replace AID as the major distributor of U.S. Government bilateral assistance. It would have as its charge the resolution of specific global development challenges jointly chosen by the Congress and the Executive, preferably concentrating on a few key global problems whose solutions would benefit large numbers of people. It would subject public sector programs and institutions to market competition. According to the proponents of the new concept, when in full operation, most of the U.S. bilateral aid budget would ultimately flow through the SDF. Discretionary resources available to other agencies such as AID would be reduced; AID could continue as a major delivery agency for assistance but would have to compete for contracts with other delivery organizations, including other government agencies, international institutions, nongovernmental organizations in both developed and developing countries, and profit-making firms. Such a complex undertaking would have to be well coordinated. This SDF proposal, and others that are more or less radical in design, deserve attention, because they aim to provide clear solutions to the deep problems now being ignored. Whatever alternatives are selected, the result must link programs to overall policy *and* couple science and technology to the missions being pursued.

THE WORLD BANK AND THE UN SYSTEM

The United States has been deeply involved with the establishment and subsequent evolution of the system of multilateral institutions that emerged after World War II. Among these institutions are the United Nations and its specialized agencies; the International Monetary Fund and the World Bank, created at the Bretton Woods Conference; and several regional development banks. The institutions have grown in number, size, and complexity, but few have reexamined their roots and mandates in the face of today's rapid change, nor conducted searching, thoughtful analysis of their future roles. The demands imposed by a rapidly changing international order require a restructuring of the existing multilateral institutional arrangements, particularly in view of the critical importance that science and technology now have in international relations. Multilateral action is likely to be more important in the future, and the vehicles will have to be more effective.

Structural Adjustment and Policy Reform

The mandates of multilateral institutions cover a variety of functions: technical assistance, grant making, investment lending, concessional resources mobilization, and policy advice and dialogue. These functions and the balance among them have evolved rapidly, particularly during the past decade. Science and technology have usually figured as subsidiary issues in the operations of the multilaterals, a situation that must be modified if the multilateral system is to respond adequately to the new development challenges in the transition to the 21st century.

The UN system and multilateral development banks are in the process of adapting to the new demands, although at different speeds and in different ways. The United Nations underwent a major managerial reorganization in early 1992. The World Bank and the IMF became universal institutions with the admission of the countries of the former Soviet Union in the spring of 1992. The European Bank for Reconstruction and Development (EBRD) was created in 1990. The June 1992 United Nations Conference on Environment and Development in Brazil has provided an extraordinary opportunity to advance the frontiers of global action for development.

The current emphasis on structural adjustment and policy reform is accompanied by a more balanced view of the relations between government intervention and market forces, a recognition of the utmost importance of environmentally sustainable development, and an acknowledgement of the key roles played by the private sector. There is also greater acceptance of contributions by nongovernmental organizations and new concern with good governance and human rights, in addition to the overriding objective of reducing poverty. The time is ripe for exploring the possible roles that multilateral institutions can play in mobilizing science and technology for development objectives.

Multilaterals, S&T, and Development

Multilateral institutions have accumulated a substantial body of experience with support for science and technology in developing countries. But it is fragmented and partial, and integration and evaluation are necessary if lessons are to be drawn for the future. In addition to providing financial resources and technical assistance through specific investment for science and technology projects, multilateral institutions include S&T components in most of their projects in other fields.

The World Bank

For example, since 1977 the World Bank has financed 31 industrial technology development projects, of which 23 projects totaling $2.05 billion have been approved since 1988.[75] Starting in the mid-1960s, the World Bank has also financed more than 150 agricultural technology projects, nearly 40 science and technology education projects, and more than 30 projects dealing with the technological aspects of transport, water supply, energy, and telecommunications.[76]

From 1981 to 1987 the World Bank invested a total of about $2.1 billion in 21 free-standing agricultural research projects and in the research components of 209 agricultural and rural development projects. The World Bank has also provided continuous support to the Consultative Group on International Agricultural Research (CGIAR), a Bank-led consortium of 40 public and private sector donors established in the early 1970s. CGIAR channels more than $230 million annually to 16 international agricultural research centers and their national counterparts in developing countries.[77]

The World Bank has also become extensively involved in information technology for development. Lending for information technology systems and telecommunications in fiscal year 1990 reached a total of $1.8 billion, a significant increase from FY 1989, when the corresponding amount was $750 million, and FY 1986, when the total was $500 million. During the past five years, about 90 percent of World Bank loans have had information technology components, rising to 93 percent in FY 1990.[78]

The United Nations

The United Nations system has also been active in the support of science and technology in developing countries. The United Nations Educational, Scientific and Cultural Organization (UNESCO) has provided technical assistance and grants to scientific research centers, higher education institutions, and S&T policy agencies; the UN Industrial Development Organization (UNIDO) has supported the development of industrial technology and extension systems; and the UN Conference on Trade and Development (UNCTAD) has provided technical assistance to government agencies in developing policies for technology transfer. In addition, the United Nations Development Programme (UNDP) has made the development of science and technology capabilities one of its priorities for the 1990s.

The Inter-American Development Bank

The Inter-American Development Bank (IDB) has had a long-standing interest in the development of higher education and science and technology in Latin America. In 1962 it provided its first eight of nearly ninety loans to universities throughout the region, and since that year the IDB has also financed twenty-five projects specifically for science and technology programs; the total value of these projects is about $1.07 billion. Other regional development banks have not been as active as the IDB in this field, but they have incorporated specific research or technology development components in their loans.[79]

U.S. Support for Multilaterals

The United States provides substantial resources to multilateral organizations with significant roles in development. In its fiscal year 1991, the World Bank group received close to $1.6 billion as the U.S. contribution to the International Development Association (IDA); IDA lends only to the poorest countries at very low interest rates. A further $150 million was contributed to the International Bank for Reconstruction and Development (IBRD), which lends to middle-income developing countries at near market rates, and to the International Finance Corporation (IFC), which lends directly to the private sector.[80]

Regional development banks, including the EBRD, received just over $400 million as United States contributions to their capital, soft loan windows, and private sector operations in FY 1991. Contributions to the International Monetary Fund during FY 1991 and FY 1992 have not been approved by Congress, despite a large request by the administration to cover the U.S. portion of the global increase approved by the IMF governing board. Through its dues and other financial transfers to special programs, the United States also provides support for S&T to the United Nations and its specialized agencies. These include the UN Development Program, the Food and Agriculture Organization, the World Health Organization, the UN Fund for Population Activities, the UN Environment Program, the World Meteorological Organization, and the International Atomic Energy Agency.

Efficiency and Effectiveness

In general, multilateral development organizations and programs are favorably regarded in the development community. The banks in particular, and more recently some other multilateral organizations, make strong efforts to attract professional staff of internationally recognized technical competence. Program management techniques flow from and capitalize on staff abilities, and rather than making technical staff members advisors to generalist program managers, multilateral agencies tend to give greater management responsibility to technical staff than bilateral agencies do.

Certain multilateral institutions also seem to have achieved greater efficiency than U.S. programs. Although quantitative comparisons are difficult, the multilateral banks generally move more resources to the developing world per professional staff member than does AID. In part this may be because bilateral grant programs are generally smaller than loans provided by multilateral banks, and because bilateral programs tend to focus on sectors where project development requires more staff time, such as education, health, population, and nutrition, rather than on lending for infrastructure, policy reform, or industrial development. The point here is simply to underscore the great importance of tracking what is efficient and what is not.

New Roles, New Directions

Nevertheless, given the vastly changed nature of the international context in the post–Cold War era and the proliferation of multilateral institutions during the past several decades, multilateral agencies could usefully review their mandates and roles for the 1990s and beyond. There are many areas of conflict and duplication between multilateral institutions, regarding, for example, responses to global challenges (environment, energy, population, migration), the provision of technical assistance, and lending for investment projects and policy reform. There is a need to foster greater complementarity and a better division of labor between the UN secretariat, the various UN specialized agencies, the Bretton Woods institutions, and the regional development banks.

■ **Greatly enhanced means must be devised for coordinating the ongoing efforts of the major donors.** Such coordination would be aimed at achieving better results, given the changing circumstances in the field. Special attention should be given to the international capacity for studies and research on the most difficult and longest-range problems in science and on technology pertinent to development: new institutions may be needed. In many of the multilateral agencies as well as the banks, a full-time senior science and technology staff (with an advisory apparatus) would be useful, as more and more activities depend upon technically alert global assessments. The increased emphasis on multilateral work and enhancing donor coordination will by no means eliminate the vital roles for bilateral programs.

Clarifying strategic directions and lines of action for multilateral institutions in the field of development in general, and especially in science and technology for development, is a high priority for the international community. Systematic and strategic conceptions of the role of S&T in development will enable multilateral institutions to anticipate the next generation of problems and to support the research to ensure better responses in the future. As these changes are made, much of the critical commentary about AID—for example, insufficient evaluation of impacts and unnecessarily centralized bureaucratic layers—must also be brought to bear on the components of the United Nations system. All of this will wring even more effectiveness from existing resources.

■ **The Task Force recommends that the United States encourage, and take a leading role in, an analysis of multilateral organizations with regard to science and technology for 21st-century development partnerships.** The multilateral review should evaluate the international capacity for studies and research on the most difficult and/or longest-range problems in science and technology pertinent to development and identify ways gaps can be addressed.

A special focus of the study should be on the shared international responsibilities for the science and technology required for balanced and sustainable international development.[81] As suggested throughout this report and in particular by examples such as tuberculosis and hunger, too little of the worldwide research and technology capability is devoted to problems of development. Over the years, new institutions have been proposed to fill this gap—for example, the U.S. Institute for Scientific and Technical Cooperation, designed in 1980, but never established. The Task Force's staff papers (see the Bibliography) have sketched many new ideas for organizing institutions of excellence, sometimes organized on a regional basis and always designed for problem-oriented programs. However, the Task Force decided to avoid making specific proposals until after general presidential and congressional reviews and the recommended reevaluation of multilateral agencies. Several initiatives relating to research on environment and development are already emerging from the Earth Summit—and from other Carnegie Commission work—involving both international centers and networks of national institutions.[82] The time is right to examine both U.S. and international mechanisms for research for development. The opportunity to assess the possible responses on the basis of regional S&T research and development institutions should not be forgone.

RESOURCES FOR DEVELOPMENT ASSISTANCE

As this report has indicated, the resources for development assistance come from both private and public sources, as profit-seeking investments and as charitable contributions. Even with much increased efficiency of expenditure, the resources being applied to global development clearly fall short of the need. The hundred billion dollars or so estimated as needed to carry out the "Agenda 21" proposed at the Earth Summit is one example of a large and increasingly well-documented bill that could be coming due in the next few decades. Individuals, companies, and governments all need to assess carefully their ability to contribute, in dollars and other ways, and to debate the evidence that might be mustered to improve the basis for planning.

SKEPTICISM AND PARADOX

The U.S. Government is at a decisive juncture with regard to provision of development assistance. Gathering taxpayers' money for development assistance is politically difficult everywhere, but it is especially difficult in the United States. Although the United States launched the modern era of foreign "aid" in the 1940s and 1950s, by now its relative economic aid effort, measured by official governmental aid as a fraction of GNP, has slipped almost to the bottom of OECD donors, though it remains substantial in absolute terms.

Prospects for U.S. Official Development Assistance (ODA) have never looked bleaker. The United States, along with other donors, is suffering from what some observers see as cumulative "aid fatigue." Aid failures and ineffectiveness are widely publicized. Moreover, in a time of recession and obvious shortcomings in improving the health and welfare of its own citizens, Americans feel that scarce resources should be aimed at domestic priorities.

The situation is paradoxical. For two generations, the Cold War hobbled the allocation of public resources to most international needs other than the military dimensions of foreign policy. Many priorities, including interest in the development of poorer countries, have trailed far behind defense claims. Suddenly, defense needs are in sharp decline. There is a new opportunity for retargeting resources, especially those related to science and technology, at genuine development priorities. Yet, now that the opportunity is at hand, many Americans and the Congress seem uninterested. There are several arguments in favor of a revival in commitment.

Arguments for Renewed Commitment

First, as the creative postwar efforts and the subsequent achievements outlined in this report suggest, development assistance genuinely aimed at development, not at other strategic or commercial objectives, has been quite effective. The chronically negative image of "assistance" is closely linked to the uneasy mix of goals that characterized it until recently. Second, agencies and individuals engaged in the process have been learning. Even more is known now about how to launch and use aid resources most effectively than was the case with the rather effective programs of a generation ago. The new tools, the new global readiness, and the remarkable power of modern communications combine to give new opportunities for high returns on investments.

Third, there is no significant competition between domestic and development-assistance claims on redeployed Cold War funds: a doubled U.S. Government development assistance program would amount to less than 5 percent of the defense budget. What America needs for *both* renewed domestic priorities and a refreshed overseas development effort is political will and a new international strategy in which development is a major component. There is budgetary room for both, and positive economic gains from each.

The wave of opportunity brought by the end of the Cold War argues against any artificial budgetary cap on development assistance. With the change in the strategic situation, some old uses of the worldwide Official Development Assistance should be curtailed and the funding assigned to new priorities. Perhaps this reallocated amount will be enough to support new efforts without increasing total ODA resources for some time. If the need arises, however, there is no reason, with good executive and legislative leadership, why the opportunities in development partnerships should not claim a modest fraction of the funding now being subtracted from defense, especially since development success would decrease the need for defense investments in the future and improve the prospects for economic growth both at home and abroad.

■ **The United States can afford to—and should—rededicate itself to a fair share of the effort on urgent development in Africa, Latin America, Asia, and the Middle East *and*, at the same time, reach out to the extraordinary opportunities in Eastern Europe and the former Soviet Union.** Resources for this expanded commitment can be found, even in constrained national budgets in the United States and elsewhere, by a shift in foreign aid budgets from military to development purposes. This shift must be accompanied by a parallel shift of expenditures from military to civil investment accounts in the countries the U.S. assists. Most important, there must be a mobilization of private financial and technical resources and of public programs outside the aid budgets.

5
CODA

SURPRISES

The 1980s brought the end of the Cold War, the emergence of AIDS, the discovery of the ozone hole, a rise in Third World debt, American initiatives for the Strategic Defense Initiative and new strategic arms reductions, and a surge of respect around the world for democratization and market mechanisms.

The next decade will have its surprises as well, and they will strongly affect the success of the approaches outlined in this report. A severe, lasting depression in the major industrialized nations, collapse of international financial institutions, and resurgence of protectionism could stop almost all progress. A rapid worldwide economic recovery, low interest, and low inflation could spur unprecedented investments regardless of deliberate government policies for international cooperation.

Breakthroughs in technology also could radically increase opportunity. For example, advances in cheap and clean energy supply, direct private global

Looking across the Hooghly River to Calcutta.
(Photograph by Christopher Warren.)

telecommunications, a second Green Revolution, or new contraceptive technology could bring revolutionary changes. On the other hand, movements that have an antiscience bias could depress the rate and distort the pattern of diffusion of humane uses of modern science and technology.

The trend today is toward more modes of international interaction: more actors, new alliances, and greater organizational variety. But could there be a contraction to a smaller set of tested modes and institutions? Will political fragmentation overtake global and regional economic integration? How will the position and influence of the United States in the world change? Will new models of "development" emerge to replace or coexist with the now dominant model based on the market economy and individual rights? What will be the critical flows of capital, information, technology, and people? Such questions must be considered as new development strategies are assessed.

SEIZING THE MOMENT

No amount of speculation will prepare us for all the surprises, good and bad. As the vocabulary of North and South, East and West, and First, Second, and Third Worlds fades, we can seek to shape a much better world, more robust in the face of both likely and unlikely futures.

America cannot afford an obsolete extension of decades' old styles of development cooperation, created as a barricade to impede political adversaries or as a convenient currency to win political allies. Fresh attention must be given to both public and private vehicles for innovation as well as to the diffusion of the science and technology that are the foundations of knowledge, skills, and, ultimately, freedom itself. Organizational invention and renewal are necessary to make use of the potential.

The sweep of events has revealed a clearing horizon, and cooperative global development can now be the course to worldwide stability and prosperity. How America will respond is not a certainty. One alternative is a new isolationism and greater protectionism. Should this come to pass, tremendous opportunities would be lost for America and for the world. The Carnegie Commission's Task Force on Development Organizations urges the United States to rededicate itself to making the world economy work for everyone and to providing for those for whom the economy currently does not. For many reasons—humanitarian, economic, and security—this is, indeed, profoundly in the national interest.

APPENDIX A
INDIVIDUALS ASSISTING THE TASK FORCE

Carol Adelman, U.S. Agency for International Development
W. Brian Arthur, Stanford University
Lutz Baehr, UN Center on Science and Technology for Development
Rose Bannigan, Board on Science and Technology for International Development
Jordan Baruch, Washington, DC
James Brasher, Carter Center
James J. Bausch, Save the Children
George Bugliarello, Polytechnic University, New York
Thomas Carroll, International Executive Service Corps
Jeffrey Clark, Carter Center
Larry Dixon, MAP International
William Drayton, Ashoka Fellowships
Walter Falcon, Stanford University
Rep. Lee Hamilton (D-Indiana)
John Hurley, MacArthur Foundation

Philip Johnston, CARE
Richard Joseph, Carter Center
Kenneth Keller, Council on Foreign Relations
Richard Krasno, Institute of International Education
Rustam Lalkaka, United Nations Fund for Science and Technology for Development
Clifford Lewis, U.S. Agency for International Development
Adetokunbo Lucas, Carnegie Corporation of New York
Patricia Barnes McConnell, Michigan State University
James Mullin, International Development Research Center, Ottawa
Henry Norman, Volunteers in Technical Assistance
Geoffrey Oldham, UN Advisory Committee on Science and Technology for Development
Thomas Pickering, U.S. Department of State
Kenneth Prewitt, Rockefeller Foundation
Ronald Roskens, U.S. Agency for International Development
Timothy Rothermel, UN Development Programme
Vernon Ruttan, University of Minnesota
Karl Schwarz, U.S. Agency for International Development
Jeffrey Schweitzer, U.S. Agency for International Development
Robert Seiple, World Vision Relief and Development
John Sewell, Overseas Development Council
Alexander Shakow, World Bank
John Swensen, Catholic Relief Services
John Temple Swing, Council on Foreign Relations
Sidney Topol, Scientific Atlanta, Inc.
Sergio Trindade, UN Center on Science and Technology for Development
Julie Da Vanzo, Rand Corporation
William Walsh, Project Hope

APPENDIX B
BIOGRAPHIES OF TASK FORCE MEMBERS AND PRINCIPAL ADVISORS AND CONSULTANTS

MEMBERS OF THE TASK FORCE

Anne L. Armstrong is a rancher in Armstrong, Texas, and chair of the board of trustees of the Center for Strategic and International Studies. From 1981 to 1990 Mrs. Armstrong chaired the President's Foreign Intelligence Advisory Board. She held cabinet rank as a Counsellor to Presidents Nixon and Ford in 1973–1974 and subsequently served as U.S. Ambassador to Great Britain. A Phi Beta Kappa graduate of Vassar, Mrs. Armstrong is a member of the board of the Smithsonian Institution and of several major corporations and is a member of the International Institute for Strategic Studies (London) and the Council on Foreign Relations. Mrs. Armstrong participated as a member of the Task Force from 1990 through July 1992 and endorses its general conclusions and recommendations.

Harvey Brooks served as dean of engineering and applied physics at Harvard University from 1957 to 1975. A solid state physicist, he worked in atomic power for the General Electric Company before joining Harvard. After his tenure as dean, Dr. Brooks became Peirce Professor of Technology and Public Policy and was one of the founders of the program in science, technology, and public policy at the Kennedy School of Government. Dr. Brooks has served on the President's Science Advisory Committee and the National Science Board and is a member of the National Academy of Sciences, the National Academy of Engineering, and the Institute of Medicine.

Former President **Jimmy Carter** (Chair) is the founder of the Atlanta-based Carter Center, a nonprofit organization that works to resolve conflict, promote democracy, preserve human rights, improve health, and fight hunger around the world. Through nonpartisan study and outreach programs, the Center has addressed the prospects for peace in the Middle East, monitored elections in Latin America, mediated conflicts in the Horn of Africa, and made significant progress in improving the health of people in developing countries. Before his election as President of the United States in 1976 Mr. Carter served as Governor of Georgia and worked as a farmer and engineer.

John R. Evans is the chair of the board of Allelix Inc., a biotechnology research and development company specializing in agricultural and health products and also chair of the Rockefeller Foundation. Dr. Evans was director of the Population, Health, and Nutrition Department of the World Bank from 1979 to 1983. Earlier he served as professor of medicine and president of the University of Toronto, where he received his MD. Dr. Evans was a Rhodes Scholar at Oxford.

Robert W. Kates is the director of the Feinstein World Hunger Program at Brown University. Professor Kates's interests include the prevalence and persistence of hunger, long-term population dynamics, sustainability of the biosphere, and natural and technological hazards. Professor Kates worked in a steel mill in Gary, Indiana, and received a PhD degree from the University of Chicago without having received an undergraduate degree. From 1962 to 1986 he was on the faculty at Clark University. From 1967 to 1969 Professor Kates directed the Bureau of Land Use Planning in Dar es Salaam, Tanzania. Professor Kates is a member of the National Academy of Sciences and in 1991 received the Presidential Medal of Science.

John P. Lewis is professor of economics and international affairs at Princeton University, where he served earlier as dean of the Wilson School of Public and International Affairs. Dr. Lewis was a member of the President's Council of Economic Advisers in 1963–1964 and director of the U.S. AID Mission to India in 1964–1969. Dr. Lewis chaired the Development Assistance Committee of the Organization for Economic Cooperation and Development from 1979 to 1981 and the World Bank and International Monetary Fund Task Force on Concessional Flows from 1982 to 1985. He is the author of numerous articles and books, including *Development Strategies Reconsidered* and *Strengthening the Poor.*

Lydia P. Makhubu is vice chancellor and professor of chemistry at the University of Swaziland. Dr. Makhubu was educated at Pius XII College in Lesotho and the University of Toronto, where she received a PhD in medicinal chemistry. She is a member of the UN Advisory Committee on Science and Technology for Development and has been active in several projects of the International Council of Scientific Unions. Along with research and university management, Dr. Makhubu maintains her work as a teacher and student advisor.

M. Peter McPherson is executive vice president of the Bank of America, where he is responsible for all negotiations of restructured sovereign debt in developing countries. From 1987 to 1989 Mr. McPherson was deputy secretary of the Department of the Treasury, focusing on trade, tax, and international issues. From 1981 to 1987 he was administrator of the Agency for International Development and chair of the Overseas Private Investment Corporation. Trained in political science and law, Mr. McPherson served as a Peace Corps volunteer in Peru in 1964–1965.

Rodney W. Nichols (Vice Chair) is chief executive officer of the New York Academy of Sciences. Mr. Nichols served as vice president and executive vice president of The Rockefeller University from 1970 to 1990, following several assignments in research and technology management within the Office of the Secretary of Defense from 1966 to 1970. Trained in applied physics at Harvard, he was one of the leaders of the U.S. delegation to the 1979 UN Conference on Science and Technology for Development. He also served on the UN Advisory Committee on Science and Technology for Development. During 1990–1992, Mr. Nichols was a scholar-in-residence with Carnegie Corporation of New York and was the principal author of *Science and Technology in U.S. International Affairs*, a report of the Carnegie Commission on Science, Technology, and Government.

Rutherford M. Poats served with the U.S. Agency for International Development from 1961 to 1970, rising from area specialist to Deputy Administrator. From 1981 to 1985 he served as Ambassador to the Organization for Economic Cooperation and Development and chair of the OECD's Development Assistance Committee. Mr. Poats worked as a reporter in China, Korea, and Japan from 1947 to 1957, becoming chief of the Tokyo bureau of UPI. He is a senior consultant to the World Bank and the author of several books, including *Technology for Developing Nations.*

Francisco R. Sagasti served from 1987 to 1991 as chief of strategic planning for the World Bank. Earlier Dr. Sagasti served as an advisor to the Peruvian Ministries of Foreign Affairs and Planning and Industry, and as a board member of an engineering design firm and professor at the Universidad del Pacífico in Lima. In 1979 he worked closely with the Secretary General in the planning of the United Nations Conference on Science and Technology for Development and later chaired the UN Advisory Committee on Science and Technology for Development. From 1973 to 1978 Dr. Sagasti led an international study of S&T policies in ten developing countries, in which more than 150 researchers in Africa, Asia, Latin America, and the Middle East participated.

George P. Shultz, Jr., served as Secretary of State from 1982 to 1989 and is now a fellow of the Hoover Institution and a professor at Stanford University. Dr. Shultz was trained in economics at Princeton and MIT, specializing in studies of labor markets, and served on the faculty at MIT and the University of Chicago. Dr. Shultz was Secretary of Labor (1969–1970), Director of the Office of Management and Budget (1970–1972), and Secretary of the Treasury (1972–1974). From 1974 to 1982 Dr. Shultz was president of Bechtel Group, Inc. Dr. Shultz served as Senior Advisor to the Task Force.

ADVISORS AND CONSULTANTS

Jesse H. Ausubel is director of studies of the Carnegie Commission on Science, Technology, and Government and a fellow in Science and Public Policy at The Rockefeller University. From 1977 to 1988 Mr. Ausubel was associated with the National Academy complex, serving as a fellow of the National Academy of Sciences, a staff officer with the National Research Council Board on Atmospheric Sciences and Climate, and director of programs for the National Academy of Engineering. He was one of the main organizers of the first UN World Climate Conference and is the author of numerous publications on technology and environment.

William H. Foege has served as the Executive Director of the Carter Center of Emory University in Atlanta since 1986. In addition to directing the Carter Center's domestic and international health programs, he oversees the Task Force for Child Survival and Development and Global 2000, Inc. Former Director of the U.S. Centers for Disease Control, Dr. Foege participated in the successful eradication of smallpox in the 1970s. Dr. Foege received an MD from the University of Washington and an MPH from Harvard University.

Victor Rabinowitch was trained as an avian ecologist, receiving his advanced degrees from the University of Wisconsin in Madison. He received a PhD in the unusual combination of zoology and international relations. Dr. Rabinowitch served as director of the National Academy of Sciences' Board on Science and Technology for International Development (1970–1981), Committee on International Security and Arms Control (1985–1987), and Office of International Affairs (1981–1990). Currently, he is Vice President for Programs for the John D. and Catherine T. MacArthur Foundation.

Susan Ueber Raymond holds a PhD in international and public health from the Johns Hopkins University. She has worked as a program manager and project officer, financial analyst, negotiator, and designer of databases, specializing in health and medical systems and international economics. She worked for the World Bank from 1978 to 1980; from 1980 to 1984 she was senior vice president at the Center for Public Resources in New York City. She has consulted extensively with the U.S. Agency for International Development, and was one of the principal authors of the 1989 AID report *Development and the National Interest.*

Maryann Roper is a science consultant at the Carter Center of Emory University in Atlanta, Georgia. She received her MD degree from the Pennsylvania State University/Hershey College of Medicine. A pediatric oncologist, Dr. Roper has served on the medical school faculties of the University of Alabama and Georgetown University. Before working at the Carter Center, Dr. Roper was Deputy Director of the National Cancer Institute at the National Institutes of Health, Bethesda, Maryland.

Walter A. Rosenblith is a former foreign secretary of the National Academy of Sciences and provost emeritus of the Massachusetts Institute of Technology. Dr. Rosenblith's major fields of interest are brain research and biophysics, science and technology in the university and society, and international science, its structure, and partners. He served as Vice President of the International Council of Scientific Unions from 1984 to 1988, and he chairs the advisory panel for the World Bank's Chinese University Development Project. Dr. Rosenblith is a member of the National Academy of Sciences, the National Academy of Engineering, and the Institute of Medicine.

Charles F. Weiss, Jr., holds the PhD degree in chemical physics and biochemistry from Harvard University. Dr. Weiss began his career researching photosynthesis at Columbia University and the University of California at Berkeley. From 1971 to 1986 he was the first Science and Technology Advisor to the World Bank, where he developed and put into practice a strategy for a major expansion of the Bank's support to science and technology. From 1986 to 1989 he taught at the University of Pennsylvania, specializing in technology management and international development. He now teaches at Princeton University. He is president of International Technology Management and Finance, Inc., a firm that advises companies and governments on strategies for technology development and environmental management.

NOTES AND REFERENCES

1. J. M. Hamilton, *Main Street America and the Third World.* Seven Locks Press, Bethesda, Maryland, 1989.

2. Alan Woods, *Development and the National Interest: U.S. Economic Assistance in the 21st Century.* AID, 1989. Based on U.S. data from the Bureau of Census dating from the early 1800s and LDC data from the World Bank World Data Tables, 1960–1986.

3. World Data Tables, 1960–1986. *World Development Report*, World Bank, 1991.

4. H. G. Ogden, *CDC and the Smallpox Crusade.* HHS publication no. CDC 87-8400. Department of Health and Human Services, Washington, DC, 1987.

5. *World Development Report*, 1979 through 1991.

6. World Data Tables, 1960–1986. *World Development Report*, World Bank, 1991.

7. Bellagio Declaration: Overcoming Hunger in the 1990s, 16 November 1989. Reprinted in *Food Policy* 15(4): 286–298.

8. Woods, *op. cit.*, note 2 above, Table I: Economic Growth.

9. *Ibid.*

10. *Human Development Report*, 1991. United Nations Development Program, p. 33.

11. Woods, *op. cit.*, note 2 above, Table I: Economic Growth.

12. *World Development Report,* 1991. Table 3: Structure of Production.

13. *Ibid.* pp. 208–209.

14. *World Development Report, 1991.* Table 14: Growth in Merchandise Trade.

15. R. C. O. Mathews *et al.*, *British Economic Growth 1856–1973*. Clarendon Press, Oxford, 1982.

16. *Handbook of International Trade and Development Statistics, 1989*, UN Conference on Trade and Development, 1990. *Direction of Trade Statistics Yearbook, 1990*, International Monetary Fund, 1991.

17. *International Trade and Statistics Yearbook, Vol. 1, 1989*. United Nations, 1991. Special Table C, "Quantum Indices of Manufactured Goods Exports."

18. United Nations Conference on Environment and Development. *Agenda 21*. United Nations, New York, 1992.

19. *Human Development Report 1991*, United Nations Development Program, p. 33. R. J. Lapham and W. Parker Maudlin, "The Effects of Family Planning on Fertility: Research Findings," in *Organization for Effective Family Planning Programs*, R. J. Lapham and G. B. Simmons (eds.), National Academy Press, Washington, DC, 1987.

20. Trudy Harpham, Tim Lusty, and Patrick Vaughan (eds). *In The Shadow of the City: Community Health and the Urban Poor.* Oxford University Press, New York, 1988, p. 47.

21. Calculated from World Bank World Data Tables, 1960–1988.

22. Woods, *op. cit.*, note 2 above.

23. *World Development Report,* 1991.

24. Fortune 500 Ranking, 1990. *Fortune*, April 22, 1991.

25. Woods, *op. cit.*, note 2 above.

26. National Research Council, *The Future of the U.S. Academic Research Enterprise*, National Academy Press, 1991. Data in 1988 constant dollars.

27. Woods, *op. cit.*, note 2, p. 20.

28. Calculated from *World Development Report, 1991*.

29. *Estimates and Projections of Urban, Rural and City Populations, 1950–2025*. United Nations, 1985, p. 12.

30. Technologies for Improving Uses of Urban Resources: Summary of a Planning Meeting. Food–Energy Nexus Program and Board on Science and Technology in Development, Washington, DC, June 20–21, 1988.

31. *Migration, Population Growth and Employment in Metropolitan Areas in Selected Developing Countries.* United Nations, 1985, p. 1.

32. National Research Council, *Population Growth and Economic Development: Policy Questions.* Working Group on Population Growth and Economic Development, D. G. Johnson, chair. National Academy Press, Washington, DC, 1986.

33. R. J. Lapham and George B. Simmons (eds.), *Organizing for Effective Family Planning Programs.* National Academy Press, Washington, DC, 1987.

34. John W. Sewell, "Foreign Aid for a New World Order." *The Washington Quarterly*, Summer 1991, p. 39.

35. *Human Development Report, op. cit.*, note 22 above, p. 23.

36. Eduardo Lachica, "World Bank Issues Warning on Debt Woes." *Wall Street Journal*, December 16, 1991, p. A9C.

37. Article 19, *Starving in Silence: A Report on Famine and Censorship.* Article 19, International Centre on Censorship, London, April 1990.

38. David Morrison, "National Security: Skewed Priorities." *National Journal*, July 6, 1991, p. 1711.

39. *Ibid.*

40. *Ibid.*

41. Leslie Gelb, "More Arms, Less Aid." *Washington Post,* May 8, 1992.

42. *1991 UNESCO Statistical Yearbook.* UNESCO, Paris.

43. Robert E. Evenson, "Comparative Evidence on Returns to Investment in National and International Research Institutions," Chapter 9, pp. 237–264 in Thomas M. Arndt, Dana C. Dalrymple, and Vernon W. Ruttan, eds., *Resource Allocation and Productivity in National and International Agricultural Research*, University of Minnesota, Minneapolis, 1977.

44. U.S. Bureau of the Census, *Statistical Abstract of the United States: 1991* (111th edition). U.S. Government Printing Office, Washington, DC, 1991.

45. Lester A. Davis, Surge in U.S. Exports Supports Economy, Employment. *Business America*, May 18, 1992, p. 27.

46. *Ibid.*
47. *Ibid.*
48. Department of Commerce data in *Louis Rukeyser's Business Almanac*, Louis Rukeyser, Editor-in-Chief, and John Cooney, Managing Editor. Simon and Schuster, New York, 1991, pp. 245–257.
49. *Ibid.*
50. Louis Rukeyser, *op. cit.*, note 48 above.
51. *Ibid.*
52. The General Agreement on Tariffs and Trade (GATT) is an ongoing intergovernmental conferencing mechanism where changes in trade rules and harmonization of trade policies are sought.
53. James Mullin, The Comprehensive Review of the Operational Activities of the United Nations. A report submitted to the Office of the Special Coordinator for Operational Activities, April 21, 1992.
54. Charles Weiss, Jr., Lessons from Eight "Reform Commissions" on the Organization of Science and Technology in U.S. Bilateral Assistance. Background paper prepared for the Task Force on Development Organizations, Carnegie Commission on Science, Technology, and Government. October 1990.
55. Rukeyser, *op. cit.*, note 48 above.
56. Calculated from Davis, *op. cit.*, note 45 above.
57. Project HOPE Sends Industry Aid to CIS. *Health Horizons*, May 1992, p. 26.
58. Business Council for Sustainable Development, Stephen Schmidheiny, chair, *Changing Course: A Global Business Perspective on Development and the Environment*, MIT Press, Cambridge, 1992.
59. Report of the Task Force on Foreign Assistance to the Committee on Foreign Affairs, U.S. House of Representatives, February 1989.
60. *Ibid.*
61. *Ibid.*
62. Susan Raymond, Foreign Assistance Legislation. Background paper prepared for the Task Force on Development Organizations, Carnegie Commission on Science, Technology, and Government, October 1990.
63. Calculated from 1992 Budget Submission to Congress of the Agency for International Development.
64. L. Hamilton, Report of the Task Force on Foreign Assistance to the Committee on Foreign Affairs, U.S. House of Representatives. February 1989.
65. See, for example, two of the Task Force working papers by Susan Raymond, "Vaccine Case Study" and "S&T for Development in the Federal Government: Organizational Sketches."
66. From Fiscal Year 1993 Congressional Budget Submission.
67. *Ibid.*
68. Staffing numbers are difficult to calculate with absolute accuracy. As has been noted by the President's Commission on the Management of AID Programs, personnel planning and statistics are not uniform or complete among the various types of personnel employed by AID. The numbers used in this report are from AID's Summary Report of Direct and Non-Direct Hire Personnel: Worldwide On-Board Employment, as of December 30, 1990.
69. Calculated from "Summary of Support and Revenue," U.S. PVOs Registered with AID, Part 1—U.S. Government Support; Part 2—Non-U.S. Government Support, FY1986.
70. Calculated from Current Technical Service Contracts and Grants Active During the Period October 1, 1987, through September 30, 1988, Office of Procurement, Agency for International Development.
71. John Kean and Allen Turner, Analysis of Institutional Sustainability Issues in *USAID: 1985–86 Project Evaluation Reports.* Agency for International Development, 1987.
72. Joseph Lieberson and Devorah Miller. A Synthesis Study of the Factors of Sustainability in A.I.D. Health Projects. Center for Development Information and Evaluation, Agency for International Development, 1989.
73. Report of the President's Commission on the Management of AID Programs, George Ferris, chair. April 1992.

74. John W. Sewell and Peter M. Storm, *Challenges and Priorities in the 1990s: An Alternative U.S. International Affairs Budget, FY1993*. Overseas Development Council, Washington, DC, 1992, pp. 43–46.

75. Joshua A. Muskin, World Bank Lending for Science and Technology, PHREE/92/51R, World Bank, Washington, DC, March 1992. F. R. Sagasti. Science and Technology Policy Research for Development: An Overview and Some Priorities from a Latin American Perspective, *Science and Public Policy*, 18(6): 379–384, 1991.

76. F. R. Sagasti, International Cooperation in a Fractured Global Order, Address at the UNESCO Colloquium on the Future of International Cooperation in Science and Technology for Development. *Futures*, May 1990, pp. 417–422.

77. Anthony Pritchard, Lending by the World Bank for Agricultural Research: Review of the Years 1981–1987. World Bank Technical Paper #118, 1990.

78. Sagasti, *op. cit.*, note 75 above.

79. Data provided by the S&T Group of the Inter-American Development Bank.

80. From Congressional Budget Submission, 1991.

81. David Dickson, *The New Politics of Science.* Chapter 4. University of Chicago Press, 1988.

82. Carnegie Commission on Science, Technology, and Government, *International Environmental Research and Assessment: Proposals for Better Organization and Decision Making.* New York, July 1992.

BIBLIOGRAPHY

BACKGROUND PAPERS

Ten background papers prepared for the Task Force on Development Organizations have been reprinted in a single volume—*The United States and Development Assistance*, Carnegie Commission on Science, Technology, and Government, New York, June 1992.

Harrell, Edgar C. "Scientific and Technological Decision-Making in Japanese Bilateral Assistance." 1989.

Mosher, David. "Three Technical Programs of the World Health Organization." 1989.

Raymond, Susan. "Science and Technology at the U.S. Agency for International Development." 1989.

Raymond, Susan. "S&T for Development in the Federal Government: Organizational Sketches." 1990.

Raymond, Susan. "USAID: Organizational Update." 1990.

Raymond, Susan. "Foreign Assistance Legislation." 1990.

Raymond, Susan. "Designing and Implementing a Multiagency Project: the Vaccine Project in India." 1990.

Weiss, Jr., Charles. "Science and Technology for Development: Lessons from Experience in Development Assistance outside the United States." 1989.

Weiss, Jr., Charles. "Scientific and Technological Decision-Making at the World Bank." 1989.

Weiss, Jr., Charles. "Lessons from Eight 'Reform Commissions' on the Organization of Science and Technology in U.S. Bilateral Assistance." 1990.

GENERAL BIBLIOGRAPHY

Achebe, C., G. Hyden, C. Magadza, and A. P. Okeyo. *Beyond Hunger in Africa.* Heineman, Nairobi, 1990.

Advisory Committee on Science and Technology for Development. "Activities of the United Nations System in Science and Technology for Development" (GOA Report). Intergovernmental Committee on Science and Technology for Development A/CN.11/91, 23 May 1989.

Article 19. *Starving in Silence: A Report on Famine and Censorship.* Article 19, International Centre on Censorship, London, April 1990.

Barrows, L. "The President's Task Force on International Development, 1969–70." From appendices, Report of the Commission on the Organization of the Government for the Conduct of Foreign Policy (the Murphy Commission Report). U.S. Government Printing Office, Washington, DC, June 1975.

Bauer, P. T. *Reality and Rhetoric: Studies in the Economics of Development.* Harvard University Press, Cambridge, Massachusetts, 1984.

Baum, W. C. *Partners against Hunger: The Consultative Group on International Agricultural Research.* World Bank, Washington, DC, 1986.

Baum, W. C., and S. M. Tolbert. *Investing in Development: Lessons of World Bank Experience.* Oxford University Press, New York, 1985.

Berg, R. J., and D. F. Gordon (eds.). *Cooperation for International Development: The United States and the Third World in the 1990s.* Lynne Rienner Publishers, Boulder, Colorado, 1989.

Broad, R., J. Cavanagh, and W. Bello. "Development: The Market Is Not Enough," *Foreign Policy* 81(Winter 91–92): 144–162.

Brooks, H., "A Critique of the Concept of Appropriate Technology." In *Appropriate Technology and Social Values: A Critical Appraisal*, F. A. Long and A. Olson (eds). Ballinger, Cambridge, Massachusetts, 1980.

Brooks, H. "The Concepts of Sustainable Development and Environmentally Sound Technology." *Advanced Technology Assessment System*, June 7 (Spring 1992), pp. 19–24.

Commission on Health Research for Development, J. R. Evans (chair). *Health Research: Essential Link to Equity in Development.* Oxford University Press, New York, 1990.

Commission on the Organization of the Government for the Conduct of Foreign Policy. Robert D. Murphy, chair. U.S. Government Printing Office, Washington, DC, June 1975.

Commission on Security and Economic Assistance. Frank Carlucci, chair. Report to the Secretary of State. U.S. Government Printing Office, Washington, DC, 1983.

Daedalus. "A World to Make." (Special issue on development.) Vol. 118, 1, 1989.

DeAngelis, M. F. "Foreign AID: The Transition from ICA to AID, 1960–61." From appendices, Commission on the Organization of Government for the Conduct of Foreign Policy. June 1975.

De Long, J. B., and B. Eichengreen. "The Marshall Plan: History's Most Successful Structural Adjustment Program." Working Paper No. 3899. National Bureau of Economic Research, Cambridge, Massachusetts, November 1991.

Department of State. Some United States Activities Using Science and Technology for Development. Pub. 8990, International Organization and Conference Series 142. Washington, DC, August 1979.

Development Assistance Committee, A. R. Love, chair. *Development Co-operation: 1990 Report.* Organization for Economic Cooperation and Development. Paris, 1990.

Development Assistance Committee, J. Wheeler, chair. *Development Co-operation: 1991 Report.* Organization for Economic Cooperation and Development. Paris, 1991.

Dickson, D. "Science and Foreign Policy: Knowledge As Imperialism." Chapter 4, in *The New Politics of Science.* The University of Chicago Press, 1988.

Franke, R., and B. Chasin. "Development without Growth: The Kerala Experiment." *Technology Review* 93(3):42–51, 1990.

Gaillard, J. "Science in the Developing World: Aid and National Policies at a Crossroad." *Ambio* 19(8):348–353, 1990.

Gardner, John W. AID and the Universities: A Report to the Administrator of the Agency for International Development. Washington, DC, 1964.

General Accounting Office. Foreign Assistance: International Resource Flows and Development Assistance to Developing Countries. GAO/NSIAD-91-25FS. Washington, DC, October 1990.

General Accounting Office. Foreign Aid: Problems and Issues Affecting Economic Assistance. GAO/NSIAD-89-6-61BR. Washington, DC, December 1988.

General Accounting Office. AID Management: Strategic Management Can Help AID Face Current and Future Challenges. GAO/NSIAD-92-100. Washington, DC, March 1992.

General Accounting Office. Foreign Assistance: A Profile of the Agency for International Development. GAO/NSIAD-92-148. Washington, DC, April 1992.

Golden, W. T. (ed.). *Worldwide Science and Technology Advice to the Highest Levels of Governments.* Pergamon, New York, 1991.

Gruebler, A., and H. Nowotny. "Towards the Fifth Kondratiev Upswing: Elements of an Emerging New Growth Phase and Possible Development Trajectories." *International Journal of Technology Management* 5(4):431–471, 1990.

Hagen, J. M. "Development Policy under Eisenhower and Kennedy." *Journal of Developing Areas* 23:1–30, October 1988.

Hamilton, L. Report of the Task Force on Foreign Assistance to the Committee on Foreign Affairs. U.S. House of Representatives, February 1989.

Hancock, G. *The Lords of Poverty: The Power, Prestige, and Corruption of the International Aid Business.* Atlantic Monthly, New York, 1989.

Hellinger, S., D. Hellinger, and F. M. O'Regan. *Aid for Just Development: Report on the Future of Foreign Assistance.* Lynne Rienner, Boulder, Colorado, 1988.

Hoben, A. "USAID: Organizational and Institutional Issues and Effectiveness." Chapter 12 in *Cooperation for International Development: The United States and the Third World in the* 1990s, R. J. Berg and D. F. Gordon (eds.), Lynne Rienner Publishers, Boulder, Colorado, 1989, pp. 253–278.

Homer-Dixon, T. F. Environmental Change and Violent Conflict. Occasional Paper 4. American Academy of Arts and Sciences, Cambridge, Massachusetts, June 1990.

Hyden, K. G. "Enabling Rural Development: Insiders and Outsiders." In *Third World Affairs 1986*. Third World Affairs, London, 1986, pp. 245–255.

International Council of Scientific Unions. *Agenda for Science for Environment and Development.* 1992.

Jepma, C. J. *The Tying of Aid.* Organization for Economic Cooperation and Development, Paris, March 1991.

Kelley, A. C. "Economic Consequences of Population Change in the Third World." *Journal of Economic Literature* XXVI:1685–1728, 1988.

Kennedy, J. V., and V. W. Ruttan. "A Reexamination of Professional and Popular Thought on Assistance for Economic Development: 1949–1952." *Journal of Developing Areas* 20:297–326, 1986.

Kates, R. (guest editor). "Overcoming hunger in the 1990s." *Food Policy.* 15(4), August 1990.

Krueger, A. O., C. Michalopolous, and V. W. Ruttan. *Aid and Development.* Johns Hopkins, Baltimore, 1989.

Lewis, J. P. "Government and National Economic Development." *Daedalus* 118(1):69–88, 1989.

Lewis, J. P. External Funding of Development-Related Research: A Survey of Some Major Donors. International Development Research Center, Ottawa, September 1987.

Long, E. J., and F. Campbell. Reflections on the Role of AID and the U.S. Universities in International Agricultural Development. Project No. 936-1406, Contract No. DPE 1406-C-00-8053. Agency for International Development, Washington, DC, 1988.

Lyman, P. "Beyond AID: Alternative Modes of Cooperation." Chapter 14 in *Cooperation for International Development: The United States and the Third World in the 1990s*, R. J. Berg and D. Gordon (eds.), pp. 302–321.

Manor, J., and E. deKadt. Organizing Development Research. Institute of Development Studies. University of Sussex, England, February 1990.

McGuire, M. M., and V. W. Ruttan. "Lost Directions: U.S. Foreign Assistance Policy Since New Directions." *Journal of Developing Areas* 24:127–180, January 1990.

Messer, E., and P. Heywood. "Trying Technology." *Food Policy* 15(4):336–345.

National Research Council. *Population Growth and Economic Development: Policy Questions.* Working Group on Population Growth and Economic Development, D.G. Johnson, chair. National Academy Press, Washington, DC, 1986.

Modelski, G., and Gardner Perry, III. "Democratization from a Long-Term Perspective." In *Diffusion of Technologies and Social Behavior*, N. Nakicenovic and A. Gruebler (eds.). Springer, Berlin, 1991, pp. 19–35.

Nichols, R. W. "For Want of a Nail: An Assessment of Prospects for the United Nations Conference on Science and Technology for Development." *Technology in Society* 1:87–106, 1979.

Osterfeld, D. "The Failures and Fallacies of Foreign Aid." *Freeman*, February 1990, pp. 61–71.

Perez, C., and L. Soete. "Catching up in Technology: Entry Barriers and Windows of Opportunity." In *Technical Change and Economic Theory*, G. Dosi, C. Freeman, R. Nelson, G. Silverberg, and L. Soete (eds.). Pinter, London, 1988, pp. 458–479.

Phoenix Group. The Convergence of Interdependence and Self-Interest: Reforms Needed in U.S. Assistance to Developing Countries. International Trade and Development Education Foundation, February 1989.

Poats, R. "Applying Science and Technology to Development in the 1990s." Organization for Economic Cooperation and Development. DCD/90/13. Paris, 11 April 1990.

President's Commission on the Management of AID Programs. Report to Congress—An Action Plan. George M. Ferris, Jr., chair. Washington, DC, April 1992.

Rondinelli, D. "Reforming US Foreign Aid Policy: Constraints on Development Assistance." *Policy Studies Journal* 18(1), Fall 1989, pp. 71–89.

Rosenberg, N., and L. E. Birdzell. *How the West Grew Rich: The Economic Transformation of the Industrial World.* Basic Books, New York, 1987.

Ruttan, V. W. "The Future of U.S. Foreign Economic Assistance." Department of Economics, University of Minnesota, 12 February 1991.

Ruttan, V. W. "Why Foreign Economic Assistance?" *Economic Development and Cultural Change* 37(2):411–424, 1989.

Sachs, J. D. (ed.). *Developing Country Debt and the World Economy.* University of Chicago Press, 1989.

Sagasti, F. R. "An Institutional Approach to National Development Planning." *Technological Forecasting and Social Change* 37:321–334, 1990.

Sagasti, F. R. "Science and Technology Policy Research for Development: An Overview and Some Priorities from a Latin American Perspective." *Science and Public Policy* 18(6):379–384, 1991.

Sagasti, F. R. International Cooperation in a Fractured Global Order. Address at the UNESCO Colloquium on the Future of International Cooperation in Science and Technology for Development. Paris, June 14–16, 1989.

Salomon, J.-J. "The Importance of Technology Management for Economic Development in Africa." *International Journal of Technology Management* 5(5):523–536, 1990.

Salomon, J.-J., and A. Lebeau. "Science, Technology, and Development." *Social Science Information* 29(4):841–858, 1990.

Sewell, J. W. "Foreign Aid for a New World Order." *Washington Quarterly* Summer 1991, pp. 35–45.

Sewell, J. W., and P. M. Storm. *United States Budget for a New World Order.* Overseas Development Council, Washington, DC, 1991.

Skolnikoff, E. B. *The Elusive Transformation: Science, Technology, and the Evolution of International Politics.* Princeton University Press, 1992.

Shear, D. "U.S. Delivery Systems for International Cooperation and Development to the Year 2000." In *Cooperation for International Development: The United States and the Third World in the 1990s*, R. J. Berg and D. F. Gordon (eds.), Lynne Rienner Publishers, Boulder, Colorado, 1989, pp. 279–301.

Smuckler, R., and R. J. Berg, with D. F. Gordon. *New Challenges/New Opportunities: U.S. Cooperation for International Growth and Development in the 1990s.* Michigan State University, Center for Advanced Study of International Development, 1988.

Solow, R. M. "How to Stop Hunger." *New York Review of Books.* 5 December 1991, pp. 22–24.

Solow, R. M. "Growth Theory and After." (Nobel Lecture.) *American Economic Review* 78:307–317, 1988.

South Commission. *The Challenge to the South.* Oxford University Press, New York, 1990.

Stanfield, R. L. "Fixing Foreign Aid." *National Journal*, May 19, 1990.

United Nations. *Measures for the Economic Development of Under-Developed Countries.* New York, United Nations, 1951.

United Nations Children's Fund (UNICEF). *The State of the World's Children 1991.*

United Nations Development Program. *Human Development Report 1990.* Oxford University Press, New York, 1990.

United Nations Development Program. *Human Development Report 1991.* Oxford University Press, New York, 1991.

UNESCO. *Science and Technology for the Future: A Fresh Look at International Cooperation in Science and Technology.* End-of-decade review of the implementation of the Vienna Programme of Action. Paris, 1989.

Weiss, C., Jr. "The World Bank's Support for Science and Technology." *Science* 227:261–265, 1985.

Williams, M. J. "U.S. Coordination of Economic and Development Cooperation Policies. Chapter 11 in *Cooperation for International Development. The United States and the Third World in the 1990s*, R. J. Berg and D. F. Gordon (eds.), Lynn Rienner Publisher, Boulder, Colorado, 1989, pp. 237–252.

World Bank. *World Development Report 1991: The Challenge of Development.* Oxford University Press, New York, 1991.

World Bank. *World Development Report 1992: Development and the Environment.* Oxford University Press, New York, 1992.

World Commission on Environment and Development. *Our Common Future.* Oxford University Press, New York, 1987.

Woods, A. Development and the National Interest: U.S. Economic Assistance into the 21st Century. A report by the Administrator, Agency for International Development, February 17, 1989.

MEMBERS OF THE CARNEGIE COMMISSION ON SCIENCE, TECHNOLOGY, AND GOVERNMENT

William T. Golden (Co-Chair)
Chairman of the Board
American Museum of Natural History

Joshua Lederberg (Co-Chair)
University Professor
Rockefeller University

David Z. Robinson (Executive Director)
Carnegie Commission on Science, Technology, and Government

Richard C. Atkinson
Chancellor
University of California, San Diego

Norman R. Augustine
Chair & Chief Executive Officer
Martin Marietta Corporation

John Brademas
President Emeritus
New York University

Lewis M. Branscomb
Albert Pratt Public Service Professor
Science, Technology, and Public Policy Program
John F. Kennedy School of Government
Harvard University

Jimmy Carter
Former President of the United States

William T. Coleman, Jr.
Attorney
O'Melveny & Myers

Sidney D. Drell
Professor and Deputy Director
Stanford Linear Accelerator Center

Daniel J. Evans
Chairman
Daniel J. Evans Associates

General Andrew J. Goodpaster (Ret.)
Chairman
Atlantic Council of The United States

Shirley M. Hufstedler
Attorney
Hufstedler, Kaus & Ettinger

Admiral B. R. Inman (Ret.)

Helene L. Kaplan
Attorney
Skadden, Arps, Slate, Meagher & Flom

Donald Kennedy
Bing Professor of Environmental Science
Institute for International Studies, and
President Emeritus
Stanford University

Charles McC. Mathias, Jr.
Attorney
Jones, Day, Reavis & Pogue

William J. Perry
Chairman & Chief Executive Officer
Technology Strategies & Alliances, Inc.

Robert M. Solow
Institute Professor
Department of Economics
Massachusetts Institute of Technology

H. Guyford Stever
Former Director
National Science Foundation

Sheila E. Widnall
Associate Provost and Abby Mauze Rockefeller Professor of Aeronautics and Astronautics
Massachusetts Institute of Technology

Jerome B. Wiesner
President Emeritus
Massachusetts Institute of Technology

MEMBERS OF THE ADVISORY COUNCIL, CARNEGIE COMMISSION ON SCIENCE, TECHNOLOGY, AND GOVERNMENT

Graham T. Allison, Jr.
Douglas Dillon Professor of Government
John F. Kennedy School of Government
Harvard University

William O. Baker
Former Chairman of the Board
AT&T Bell Telephone Laboratories

Harvey Brooks
Professor Emeritus of Technology and Public Policy
Harvard University

Harold Brown
Counselor
Center for Strategic and International Studies

James M. Cannon
Consultant
The Eisenhower Centennial Foundation

Ashton B. Carter
Director
Center for Science and International Affairs
Harvard University

Richard F. Celeste
Former Governor
State of Ohio

Lawton Chiles
Governor
State of Florida

Theodore Cooper
Chairman & Chief Executive Officer
The Upjohn Company

Douglas M. Costle
Former Administrator
U.S. Environmental Protection Agency

Eugene H. Cota-Robles
Special Assistant to the Director
National Science Foundation

William Drayton
President
Ashoka Innovators for the Public

Thomas Ehrlich
President
Indiana University

Stuart E. Eizenstat
Attorney
Powell, Goldstein, Frazer & Murphy

Gerald R. Ford
Former President of the United States

Ralph E. Gomory
President
Alfred P. Sloan Foundation

The Reverend Theodore M. Hesburgh
President Emeritus
University of Notre Dame

Walter E. Massey
Director
National Science Foundation

Rodney W. Nichols
Chief Executive Officer
New York Academy of Sciences

David Packard
Chairman of the Board
Hewlett-Packard Company

Lewis F. Powell, Jr.*
Associate Justice (Ret.)
Supreme Court of the United States

Charles W. Powers
Managing Senior Partner
Resources for Responsible Management

James B. Reston
Senior Columnist
New York Times

Alice M. Rivlin
Senior Fellow
Economics Department
Brookings Institution

Oscar M. Ruebhausen
Retired Presiding Partner
Debevoise & Plimpton

Jonas Salk
Founding Director
Salk Institute for Biological Studies

Maxine F. Singer
President
Carnegie Institution of Washington

Dick Thornburgh
Undersecretary General
Department of Administration and Management
United Nations

Admiral James D. Watkins (Ret.)**
Former Chief of Naval Operations

Herbert F. York
Director Emeritus
Institute on Global Conflict and Cooperation
University of California, San Diego

Charles A. Zraket
Trustee
The MITRE Corporation

* Through April 1990
** Through January 1989

INTERNATIONAL STEERING GROUP OF THE CARNEGIE COMMISSION ON SCIENCE, TECHNOLOGY, AND GOVERNMENT

Harvey Brooks
Harvard University

Rodney W. Nichols (Chair)
New York Academy of Sciences

Victor Rabinowitch
John D. & Catherine T. MacArthur Foundation

Walter A. Rosenblith
Massachusetts Institute of Technology

Jesse H. Ausubel (rapporteur)
Carnegie Commission on Science, Technology, and Government

TASK FORCE ON DEVELOPMENT ORGANIZATIONS

President Jimmy Carter (Chair)
Rodney W. Nichols (Vice Chair)
Anne L. Armstrong*
Harvey Brooks
John R. Evans
Robert W. Kates
John P. Lewis
Lydia P. Makhubu
M. Peter McPherson
Rutherford M. Poats
Francisco R. Sagasti

George P. Shultz (senior advisor)

* Participated as a member of the Task Force from 1990 through July 1992 and endorses the general conclusions and recommendations of the report.

By Dona Herweck Rice
Illustrated by Linda Silvestri

Publishing Credits

Rachelle Cracchiolo, M.S.Ed., *Publisher*
Aubrie Nielsen, M.S.Ed., *EVP of Content Development*
Emily R. Smith, M.A.Ed., *VP of Content Development*
Véronique Bos, *Creative Director*
Dani Neiley, *Associate Editor*
Kevin Pham, *Graphic Designer*

Image Credits

Illustrated by Linda Silvestri

Library of Congress Cataloging-in-Publication Data

Names: Rice, Dona, author. | Silvestri, Linda, illustrator.
Title: Let's talk / by Dona Herweck Rice ; illustrated by Linda Silvestri. Other titles: Let us talk
Description: Huntington Beach, CA : Teacher Created Materials, [2022] | Audience: Grades 2-3. | Summary: ""What happens when all the parts of the body stop cooperating? Well, Coach B is finding out. She's the brains of the outfit! But Coach finds herself stumbling along until she remembers what it really means to be a team""-- Provided by publisher.
Identifiers: LCCN 2021052991 (print) | LCCN 2021052992 (ebook) | ISBN 9781087601953 (paperback) | ISBN 9781087632018 (ebook)
Subjects: LCSH: Readers (Primary) | LCGFT: Readers (Publications)
Classification: LCC PE1119.2 .R5358 2022 (print) | LCC PE1119.2 (ebook) | DDC 428.6/2--dc23/eng/20211118
LC record available at https://lccn.loc.gov/2021052991
LC ebook record available at https://lccn.loc.gov/2021052992"

TCM Teacher Created Materials
5482 Argosy Avenue
Huntington Beach, CA 92649
www.tcmpub.com

ISBN 978-1-0876-0195-3

Printed in Malaysia. THU001.46774

Table of Contents

Chapter One

The Brains of the Outfit

It had been a very long day, and Coach B was tired—bone-tired. She had been up since…well, forever. Brainiacs like Coach B never really slept. That was the job, and she didn't mind it in the least. Staying active is what brains do. But she *did* mind the chaos.

She was used to everyone listening to her directions and just *doing* them. She expected no arguments and no fuss. It had always been that way. But today? The whole system had gone haywire! The left hand didn't know what the right hand was doing—literally.

Coach B was still shaking her head over the hullabaloo of this day. It was a tangled mess, and she had no idea what to do. So, she did the one thing she was used to doing: she rallied the troops.

"Listen up!" Coach started to shout. "It's about time you all worked something out. This kind of chaos is no good for me. Come on and work together!" said she.

Coach B looked at the team, expecting a quick reply and a "sorry" or two. She was sure that everyone would step in line and get back to business. But nothing, *nada*, crickets. Her plea was met with silence. Everyone had a mind of their own, it seemed.

Suddenly, the silence was broken with a loud, stomping echo coming from below. Looking down, Coach B caught a glimpse of a wiggle, a kick, and another stomp. Then, she heard a loud, "Hey!"

What in the world was going on?

Chapter Two

Not Standing for It

Coach B cleared her throat and asked, "Well, do you have something to say?"

"You bet I do," came the emphatic answer. Then, the angry voice continued.

"Who do you think you are to push us about? I don't have to stand here while you shout. I won't play along or get my toes in a bunch!" the left foot jeered with a kick and a punch.

Coach B was shocked and dismayed. No one on the team had ever talked to her this way!

“Hey, I am your coach, and that is a fact.
Don’t you remember that we made a pact?
I give the directions, and you do the deed.”
Then growling, Coach added, “Just follow my lead!”

That ought to do it, Coach B thought to herself. *They’ll fall in line now.* But a flutter from below caught her eye. *Now what’s the problem?* she wondered.

Looking down, Coach B saw the foot trying to draw attention to the hand. “This doesn’t look promising,” she sighed.

"That's right! I see you looking down at me.
I'm not going to take it anymore, you see?"
Then, the right hand gave their hair a toss.
"I'm handy myself and *don't* need a boss."

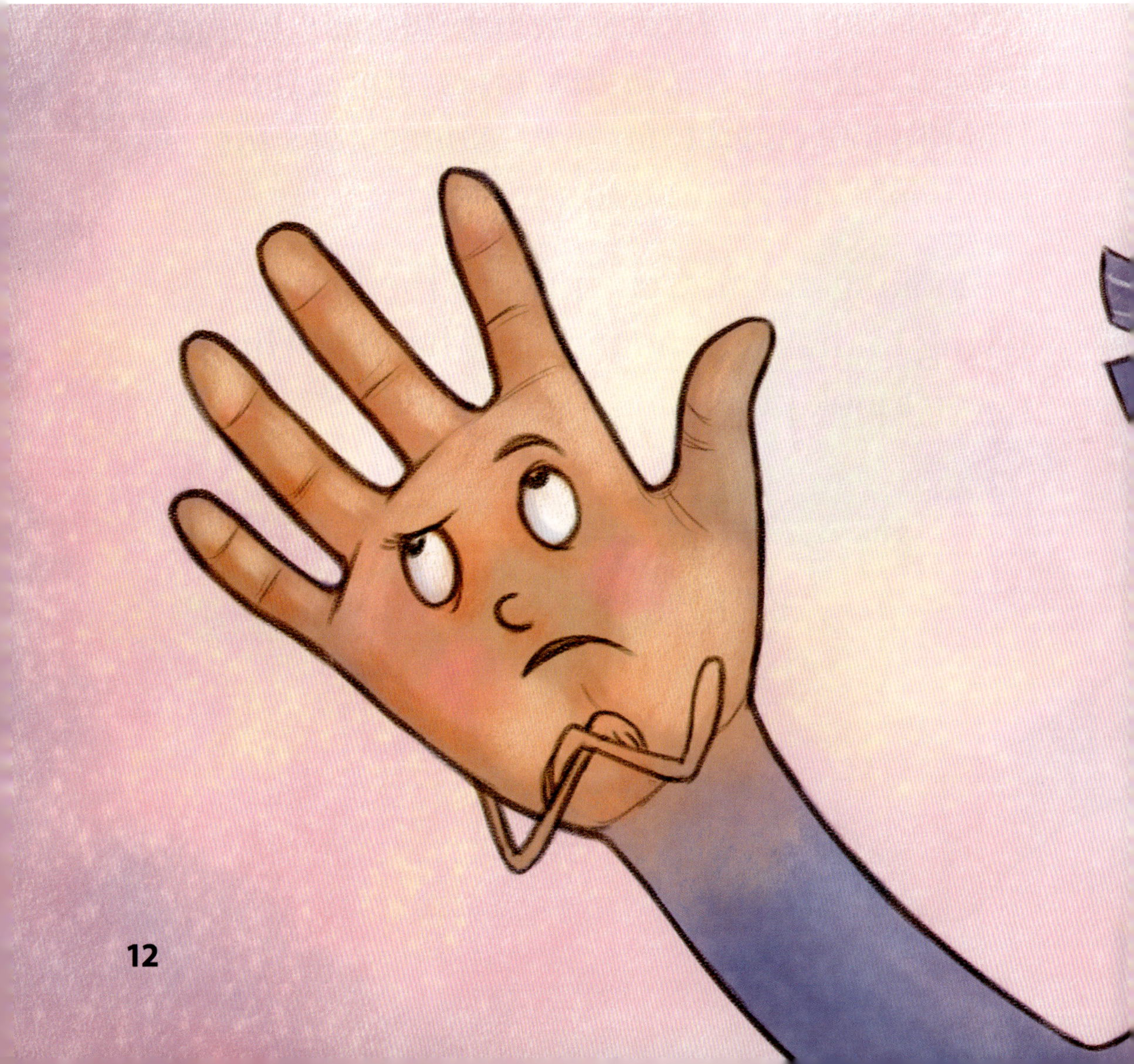

Coach B would have slapped the left hand to their forehead, but it wasn't listening either. "What in the world is happening here?" Coach asked, puzzled. "And whatever can I do about it?"

Chapter Three

Out of Joint

Coach B thought it couldn't get any worse. Then, she felt a rumbling and a quake through the whole body. Things were haywire before, but that was nothing compared to this.

The body started leaning sideways to the right. The left arm and leg began to flap wildly north and south. The right ankle spun in crazy circles, and the left knee bobbed up and down.

"What are you doing?" cried Coach with alarm. "All this mischief is causing us harm! This is no way to act! This is no way to be! Pull yourselves together, and listen to me!"

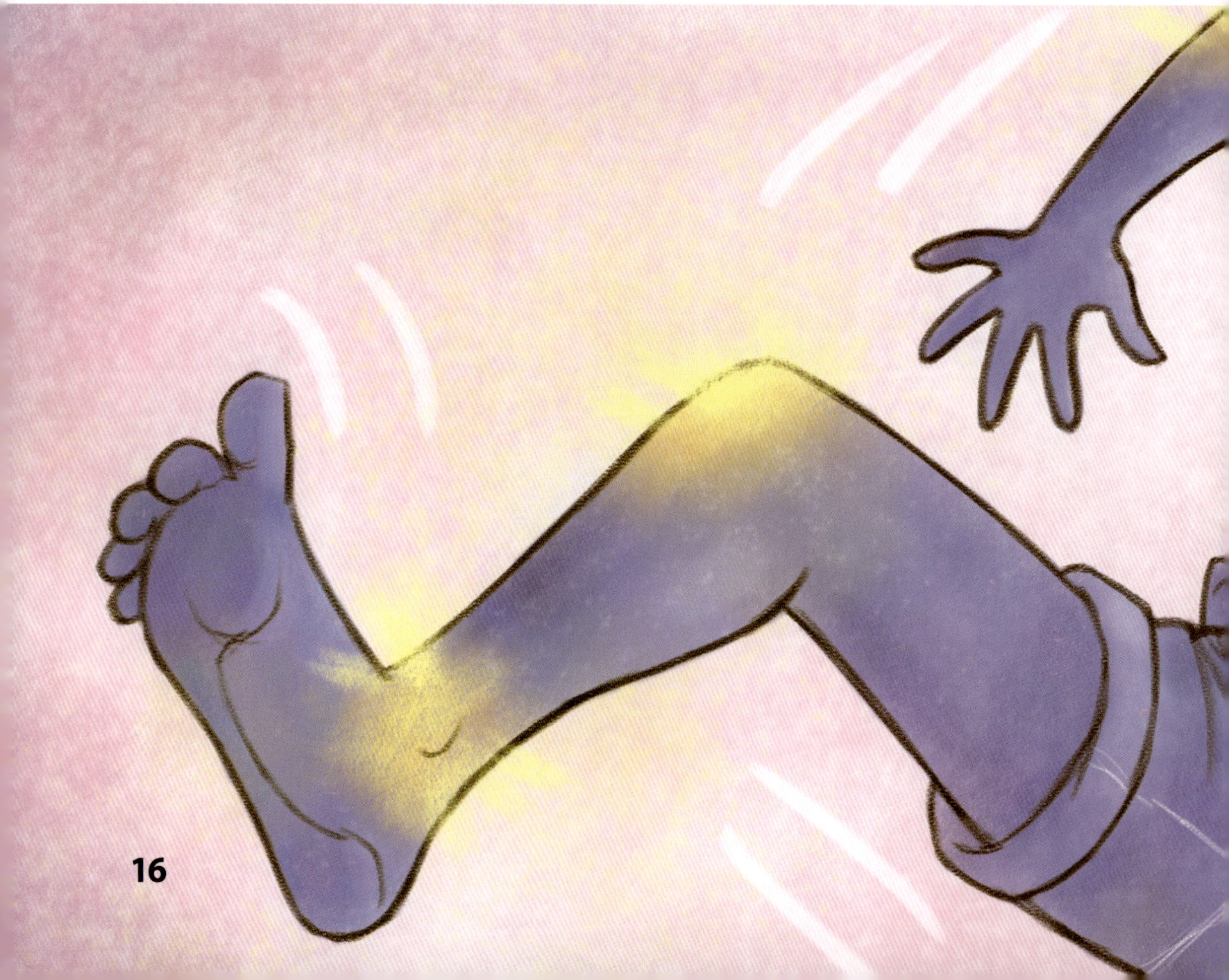

Coach waited, but there was no answer—only a lot of creaking from her right and her left. Coach looked down to see a circus of flapping and fluttering. The elbows were opening and closing at random. The knees bounced this way and that. The ankles flicked like crazy!

“Well? Spit it out, you elbows, ankles, and knees! What do you have to say for yourselves?” Coach B whooped.

“We’re tired of doing whatever you want. And being your personal restaurant—where you place the orders and we get it done.”

Then, the left knee added, “It just isn’t fun!”

Coach B wrinkled her brow and stared at the joints. Then sighing, she said, “It’s hopeless!”

Chapter Four

Teamwork

"A-hem," coughed a voice from somewhere nearby. Coach looked but couldn't see where the voice was coming from, until it spoke again.

"Excuse me," called the voice, soft and pink. "This really isn't as bad as you think. They just want to work together with you. Now, isn't that something a coach should do?"

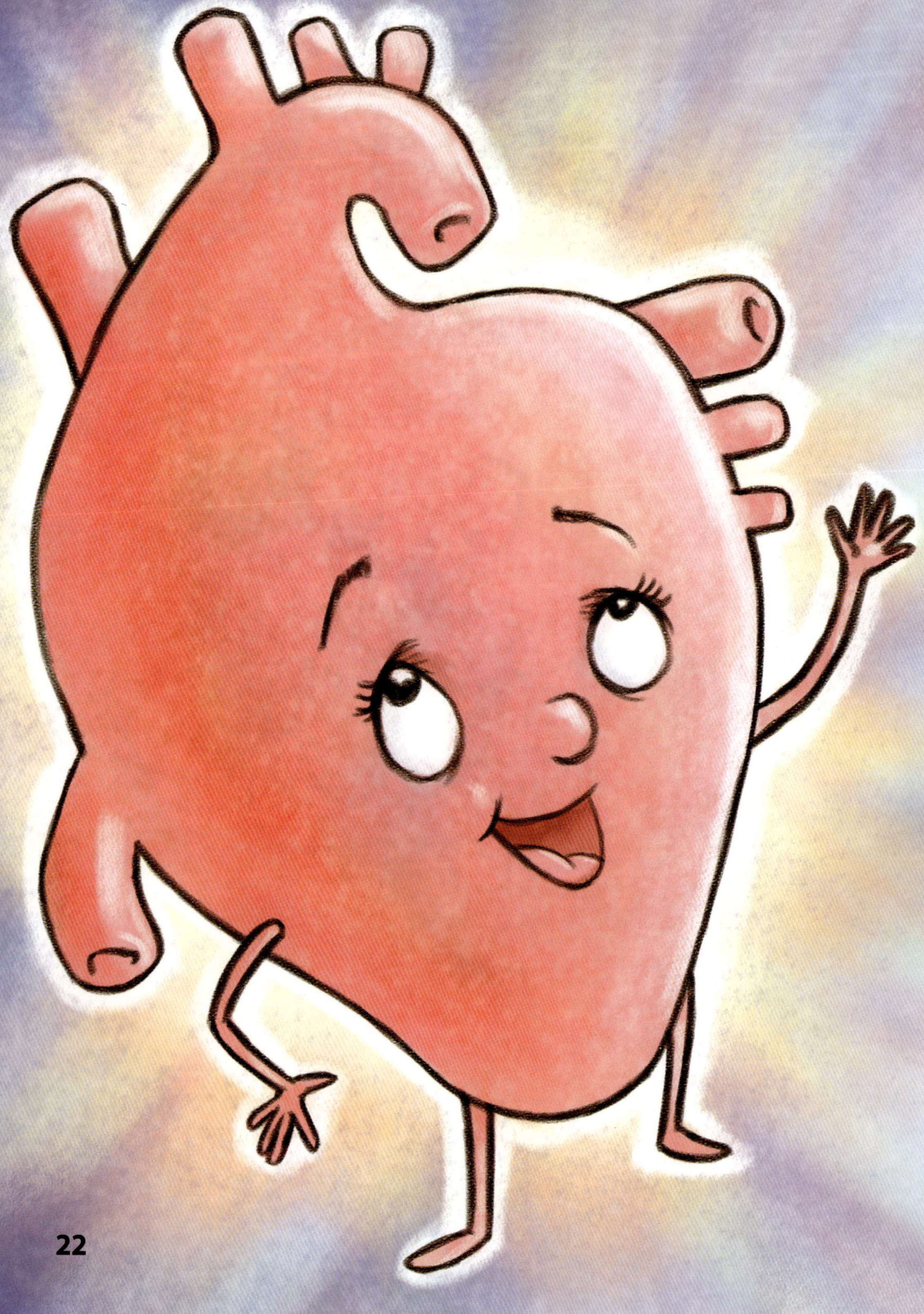

Coach B looked into the darkness. Then, she heard a steady *thump-thump, thump-thump, thump-thump.*

She shifted her gaze directly below, and there she saw the strong, constant, and reliable heart, beating in time. The heart began to glow softly and looked right back at Coach B with gentle kindness in her eyes.

"Oh, it's you," Coach said.

"Yes, it's me, dear Coach," the heart started to say. "And we don't have to go on this way. How about trying teamwork from this moment on? I'm quite certain we can all get along."

Coach turned a little pink. She was embarrassed that she, a coach, forgot the importance of teamwork. But then she shook her head and gave herself a good talking to.

"That's it!" Coach B cried. "How could I forget? Working together is the most important thing yet! And it's not just the players but the coach who must, too." Then Coach added, "Hey, team, I'm ready to listen to you!"

"Hooray!" the team cheered. "That's all that we need. We're in this together, no matter the deed. Just remember, Coach, we all have a part. And nothing matters if we don't play with heart."

And all of a sudden, without fuss or bother, all the parts were in sync with one another. Coach, as their leader, was part of the team. And together, everything worked like a dream.

Together, they worked and they danced and they swayed. They slept and they dreamed and they ran and they played.

They all worked together; they each did their part. And they never forgot that a body needs heart!

About Us

The Author

Dona Herweck Rice is very grateful when all the parts of her body work together to help her write a good book. Some days, it definitely doesn't feel like they are getting along! But after a heart-to-heart talk and a pat on the back, they usually come around. Dona lives and works in California, where she talks to herself *nearly* every day. She is happy that her family doesn't seem to mind.

The Illustrator

Linda Silvestri is an illustrator living in Southern California with two "helpful" kitty interns and her lovely husband Tom. When not prying an intern or two off of her drawing pad, you'll find Linda in her studio, busy bringing characters to life for kids of all ages.